A **VISION** FOR THE
INTERNATIONAL POLAR YEAR 2007–2008

U.S. National Committee for the International Polar Year 2007-2008
Polar Research Board
Division on Earth and Life Studies

NATIONAL RESEARCH COUNCIL
OF THE NATIONAL ACADEMIES

THE NATIONAL ACADEMIES PRESS
Washington, D.C.
www.nap.edu

THE NATIONAL ACADEMIES PRESS   500 Fifth Street, N.W.   Washington, DC 20001

NOTICE: The project that is the subject of this report was approved by the Governing Board of the National Research Council, whose members are drawn from the councils of the National Academy of Sciences, the National Academy of Engineering, and the Institute of Medicine. The members of the committee responsible for the report were chosen for their special competences and with regard for appropriate balance.

Support for this project was provided with institutional funds from the National Academies.

International Standard Book Number 0-309-09212-4 (Book)
International Standard Book Number 0-309-53203-5 (PDF)

Cover image: Glacier edge. © 1995. PhotoDisc Inc. All rights reserved. Images © 1995 Tony Ise.

Additional copies of this report are available from the National Academies Press, 500 Fifth Street, N.W., Lockbox 285, Washington, DC 20055; (800) 624-6242 or (202) 334-3313 (in the Washington metropolitan area); Internet, http://www.nap.edu.

THE NATIONAL ACADEMIES
Advisers to the Nation on Science, Engineering, and Medicine

The **National Academy of Sciences** is a private, nonprofit, self-perpetuating society of distinguished scholars engaged in scientific and engineering research, dedicated to the furtherance of science and technology and to their use for the general welfare. Upon the authority of the charter granted to it by the Congress in 1863, the Academy has a mandate that requires it to advise the federal government on scientific and technical matters. Dr. Bruce M. Alberts is president of the National Academy of Sciences.

The **National Academy of Engineering** was established in 1964, under the charter of the National Academy of Sciences, as a parallel organization of outstanding engineers. It is autonomous in its administration and in the selection of its members, sharing with the National Academy of Sciences the responsibility for advising the federal government. The National Academy of Engineering also sponsors engineering programs aimed at meeting national needs, encourages education and research, and recognizes the superior achievements of engineers. Dr. Wm. A. Wulf is president of the National Academy of Engineering.

The **Institute of Medicine** was established in 1970 by the National Academy of Sciences to secure the services of eminent members of appropriate professions in the examination of policy matters pertaining to the health of the public. The Institute acts under the responsibility given to the National Academy of Sciences by its congressional charter to be an adviser to the federal government and, upon its own initiative, to identify issues of medical care, research, and education. Dr. Harvey V. Fineberg is president of the Institute of Medicine.

The **National Research Council** was organized by the National Academy of Sciences in 1916 to associate the broad community of science and technology with the Academy's purposes of furthering knowledge and advising the federal government. Functioning in accordance with general policies determined by the Academy, the Council has become the principal operating agency of both the National Academy of Sciences and the National Academy of Engineering in providing services to the government, the public, and the scientific and engineering communities. The Council is administered jointly by both Academies and the Institute of Medicine. Dr. Bruce M. Alberts and Dr. Wm. A. Wulf are chair and vice chair, respectively, of the National Research Council.

www.national-academies.org

U.S. NATIONAL COMMITTEE FOR THE INTERNATIONAL POLAR YEAR 2007-2008

MARY ALBERT (*Chair*), Cold Regions Research and Engineering Laboratory, Hanover, New Hampshire
ROBERT BINDSCHADLER, National Aeronautics and Space Administration, Greenbelt, Maryland
CECILIA BITZ, University of Washington, Seattle
JERRY BOWEN, CBS News, Los Angeles, California
DAVID BROMWICH, The Ohio State University, Columbus
RICHARD GLENN, Arctic Slope Regional Corporation, Barrow, Alaska
JACQUELINE GREBMEIER, University of Tennessee, Knoxville
JOHN KELLEY, University of Alaska, Fairbanks
IGOR KRUPNIK, Smithsonian Institution, Washington, D.C.
LOUIS LANZEROTTI, Bell Laboratories, Lucent Technologies, Murray Hill, New Jersey
PETER SCHLOSSER, Lamont-Doherty Earth Observatory, Palisades, New York
PHILIP M. SMITH, McGeary & Smith, Santa Fe, New Mexico
GEORGE SOMERO, Stanford University, Pacific Grove, California
CRISTINA TAKACS-VESBACH, University of New Mexico, Albuquerque
GUNTER WELLER, University of Alaska, Fairbanks
DOUGLAS WIENS, Washington University, St. Louis, Missouri

Ex-Officio Members

MAHLON C. KENNICUTT II, Texas A&M University, College Station
ROBIN BELL, Lamont-Doherty Earth Observatory of Columbia University, Palisades, New York
PATRICK WEBBER, Michigan State University, East Lansing
TERRY WILSON, The Ohio State University, Columbus

NRC Staff

SHELDON DROBOT, Study Director, Polar Research Board
CHRIS ELFRING, Director, Polar Research Board
KRISTEN AVERYT, Christine Mirzayan Intern, Polar Research Board
SARAH CAPOTE, Project Assistant, Ocean Studies Board
RACHAEL SHIFLETT, Senior Project Assistant, Polar Research Board

Preface

The research ideas jointly voiced by many scientists from many nations for the International Polar Year (IPY) 2007-2008 hold potential discoveries that are significant to all inhabitants of this planet. The large-scale environmental changes currently observed in the polar regions are significant, accelerating, and globally connected. They are unlike any in recorded history, yet we do not know how or why they are occurring. Polar regions hold unique information on Earth's past climate history, for both the recent and the distant past. They play key roles in many of Earth's linked systems, from the center of the Earth to the center of the Sun. Exploration of little-known realms and processes facilitates discoveries on unanswered questions that span all disciplines.

The IPY 2007-2008 follows a tradition of international endeavors. More than a century ago, in 1882-1883, scientists around the world united in an ambitious effort to explore and conduct scientific research in the polar regions. This effort set a precedent for international science cooperation: we can achieve more, at greater efficiency, if we work together. Fifty years later, a second IPY in 1932-1933 led to major scientific advances. In 1957-1958 the science community coalesced around the most ambitious effort to date: the International Geophysical Year (IGY), where 67 participating nations left an amazing legacy of discoveries and technological accomplishments that still affect our lives today. Nearly fifty years have passed since the IGY and once again scientists around the world are eager to organize an international science campaign that expands the boundaries of our understanding of the polar regions and their key roles in the Earth's linked systems. This effort is focused on the polar regions because environmental changes currently observed there are significant, accelerating, and globally connected; the polar regions hold unique information on Earth's past climate history; they are growing in economic and geopolitical importance; the harsh conditions and remoteness have hampered scientific exploration in comparison with the midlatitudes and tropics; and the polar regions are a unique vantage point for many terrestrial and solar studies.

Planning for IPY 2007-2008 has evolved in a bottom-up fashion, with scientists in many different countries coming together to create a vision for what the next IPY might accomplish. The committee owes its thanks to many scientists in a number of countries, including the United States, who led the informal discussions that catalyzed IPY planning. The effort moved from brainstorming to official endorsement by the International Council for Science (ICSU), the World Meteorological Organization, and numerous other organizations in barely a year so that now, in the spring of 2004, nations around the globe are making concrete plans to participate.

In the United States, planning has evolved quickly. In the winter of 2002 the Polar Research Board (PRB) of the National Academies held a one-day workshop on the IPY that involved both U.S. and international scientists. The chair of the PRB, Dr. Robin Bell, began working closely with the director of the British Antarctic Survey, Dr. Christopher Rapley, to set in motion a chain of discussions that gathered ever-growing scientific support. Under their leadership, the ICSU established an IPY Planning Group in the fall of 2003. This group created a preliminary report to ICSU that led to official endorsement of the IPY and set in motion the formation of national committees devoted to IPY planning in countries around the world.

To coordinate the U.S. scientific community's efforts in identifying potential con-tributions to IPY and to provide a means for interaction with the ICSU IPY Planning Group, the PRB formed the U.S. National Committee for the International Polar Year in the summer of 2003. Everyone on this committee has generously volunteered their time to attend planning meetings, give presentations at professional meetings, compile information from the community, and write this report. Committee members have nurtured discussions with federal departments and agencies conducting or sponsoring research, recognizing that the success of the IPY will hinge on productive dialogue between national and international research communities and organizations. This bottom-up approach has resulted in a broad base of support both within the scientific communities and in the agencies for the next IPY. The committee's website at http://us-ipy.org is updated regularly to announce upcoming events and to make oppor-tunities for input readily available.

This report reflects a vision for U.S. participation in the IPY 2007-2008. It articulates a framework for the science ideas submitted by many individuals and research communities in the United States for the next IPY. These ideas are not being fully addressed by current research programs. Our intent is not to replace currently funded research but rather to lay the foundations to elevate future discoveries to a new level. The outcomes from research foundations laid in 2007-2008 hold the potential to guide important decisions for society in the twenty-first century. The committee wrote this report to convey the members' thinking to decision makers and the public, so that they can become engaged and excited about the possibilities. In addition it is hoped that many people—teachers, scout leaders, museum directors, filmmakers, journalists, parents, and students—will start to think about how they might become involved. The committee envisions that IPY 2007-2008 will initiate a new era in polar science by establishing the ongoing intellectual commitment, international research programs, and observation systems needed to fully understand the polar regions and their key roles in the global system.

I thank PRB Director Chris Elfring for initiating this study and the National Academies for supplying funding. Study Director Sheldon Drobot provided tireless support and enthusiasm; intern Kirsten Averyt inspired us to speak to the next generation; and staff members Ann Carlisle, Jodi Bachim, and Rachael Shiflett provided excellent assistance. Again, I express my deep appreciation to the scientists on the committee who are giving their time and energy to plan this important activity. I look forward to increasing interaction with my international colleagues, as the ideas submitted within our nations are shared among nations and mechanisms are established to move us from planning to implementation. I believe we will leave behind a legacy of accomplishments worthy of the traditions set by previous IPYs and the IGY.

Mary Albert, *Chair*
U.S. National Committee for the
International Polar Year 2007-2008

Acknowledgments

This report has been reviewed in draft form by individuals chosen for their diverse perspectives and technical expertise, in accordance with procedures approved by the National Research Council's Report Review Committee. The purpose of this independent review is to provide candid and critical comments that will assist the institution in making its published report as sound as possible and to ensure that the report meets institutional standards for objectivity, evidence, and responsiveness to the study charge. The review comments and draft manuscript remain confidential to protect the integrity of the deliberative process. We wish to thank the following individuals for their review of this report:

Richard Alley, Pennsylvania State University
Sridhar Anandakrishnan, Pennsylvania State University
Frank Carsey, NASA/Jet Propulsion Laboratory
Bert Boyer, University of Alaska-Fairbanks
Lou Codispoti, University of Maryland
Robert Corell, American Meteorological Society and Arctic Climate Impact Assessment
Vladimir Papitashvili, University of Michigan

Although the reviewers listed above have provided constructive comments and suggestions, they were not asked to endorse the report's conclusions or recommendations, nor did they see the final draft of the report before its release. The review of this report was overseen by Deborah Meese, U.S. Army Engineer Research and Development Center Cold Regions Research and Engineering Lab. Appointed by the National Research Council, she was responsible for making certain that an independent examination of this report was carried out in accordance with institutional procedures and that all review comments were carefully considered. Responsibility for the final content of this report rests entirely with the authoring committee and the institution.

The committee extends a special acknowledgment to Patricia McAdams for interviewing polar scientists and bringing their stories to life. Ms. McAdams also provided much of the text for the sidebars in this report.

Contents

Executive Summary

Environmental changes currently witnessed in the polar regions are vivid and in many cases greater than changes observed in the midlatitudes or tropics. The Arctic ice cover is decreasing in extent and area; some ice shelves in Antarctica are retreating and thinning; glaciers across the globe are shrinking; ecosystems are changing; Alaskan villages are being moved to higher ground in response to coastal erosion; and permafrost thawing is causing the collapse of roads and buildings. We must understand these changes in the context of past changes in order to make informed choices for the future. Yet we do not understand how or why many of the changes are occurring. Exploration of new scientific frontiers in the polar regions may help scientists better understand Earth's environment and will also lead to new discoveries, insights, and theories potentially important to all people.

In effect, the polar regions are central to many of the key science issues of our time. To better understand these and other questions, nations around the world are making plans to participate in International Polar Year (IPY) 2007-2008. At its most fundamental level, IPY 2007-2008 is envisioned to be an intense, coordinated campaign of polar observations, research, and analysis that will be multidisciplinary in scope and international in participation. IPY 2007-2008 is an ambitious program following in the footsteps of some historic past campaigns. During previous IPYs (1882-1883 and 1932-1933) and the International Geophysical Year (IGY) in 1957-1958, which was held on the 25th anniversary of the previous IPY, unprecedented exploration of Earth and space led to discoveries in many fields of science that fundamentally changed how we view the polar regions and their global linkages (see Chapter 1). In its turn, IPY 2007-2008 will benefit society by exploring new frontiers and increasing our understanding of the key roles of the polar regions in globally linked systems.

This IPY will be far more than an anniversary celebration of the IGY or previous IPYs: it will be a watershed event and will use today's powerful research tools to better understand the key roles of the polar regions in global processes. Automatic observatories, satellite-based remote sensing, autonomous vehicles, the Internet, and genomics are just a few of the innovative approaches for studying previously inaccessible realms. IPY 2007-2008 will be

1

fundamentally broader than the IGY and past IPYs because it will explicitly incorporate multidisciplinary and interdisciplinary studies, including biological, ecological,
and social science elements. Such a program will not only add to our scientific understanding but also it will result in a world community with shared ownership in
the results.

IPY 2007-2008 will provide a framework and impetus to undertake projects that
normally could not be achieved by any single nation. It will allow us to think beyond
traditional borders—whether national borders or disciplinary constraints—toward a
new level of integrated, cooperative science. A coordinated international approach
maximizes both impact and cost effectiveness, and the international collaborations
started today will build relationships and understanding that will bring long-term
benefits. Within this context, IPY 2007-2008 will seek to galvanize new and innovative
observations and research while at the same time building on and enhancing existing
relevant initiatives. It will serve as a mechanism to attract and develop a new generation
of scientists and engineers with the versatility to tackle complex global issues. In
addition, there is clearly an opportunity to organize an exciting range of educational
and outreach activities designed to excite and engage the public, with a presence in
classrooms around the world and in the media in varied and innovative formats.

Enthusiasm for IPY 2007-2008 is strong and growing. In about one year the
science community has progressed from its earliest discussions of possibilities for new

International Polar Year 2007-2008 will be an intense, internationally coordinated research campaign. It
will include an exciting array of opportunities for students and the public to both appreciate the beauty of
polar environments and better understand their importance in the global system. SOURCE: Michael Van
Woert, NOAA.

international science endeavors to serious planning of what IPY 2007-2008 might accomplish and what resources are needed. More than 25 nations have formally declared their intent to participate, and many more have discussions in process. Here in the United States, scientists have been presenting talks and holding open forums at professional meetings and using an interactive Website to brainstorm ideas where U.S. leadership might ensure significant contributions. A call to the science community for ideas about what science themes to pursue brought forward hundreds of ideas, and this input has been crucial in the IPY planning and in the preparation of this report.

The U.S. National Committee for the International Polar Year 2007-2008 was formed by the Polar Research Board of the National Academies to articulate a vision for U.S. participation in IPY 2007-2008 in coordination with and on behalf of our nation's science community. The committee has worked closely with the U.S. science community using a variety of mechanisms, and it has worked with our international colleagues, especially the International Council for Science's IPY 2007-2008 Planning Group, to identify important science themes and develop the detailed information needed to implement its many contributing activities. This report is a record of the committee's deliberations. Chapter 1 outlines the rationale for organizing IPY 2007-2008. Chapter 2 provides an overview of the scientific challenges that lay the foundation for IPY activities. Chapters 3 and 4 look in depth at the proposed main science themes identified here in the United States, which are understanding change in the polar regions and exploring new scientific frontiers. In Chapter 5 the focus is on the technology needed to support innovative observations. Chapter 6 sets the stage for discussions on capitalizing on IPY 2007-2008 to increase public understanding of the polar regions and science more broadly. Finally, Chapter 7 presents a detailed list of the actions needed to implement IPY 2007-2008. These recommendations, which are supported by discussions in the main chapters, are as follows:

Recommendation 1: The U.S. science community and agencies should use the International Polar Year to initiate a sustained effort aimed at assessing large-scale environmental change and variability in the polar regions.

• *Provide a comprehensive assessment of polar environmental changes through studies of the past environment and the creation of baseline datasets and long-term measurements for future investigations.*

Environmental changes currently observed in the polar regions are unprecedented in times of modern observation. Studies investigating natural environmental variability, human influence on our planet, and global teleconnections will help in understanding mechanisms of rapid climate change and in developing models suitable for forecasting changes that will occur in the twenty-first century. This effort will need to be sustained after IPY 2007-2008.

• *Encourage interdisciplinary studies and the development of models that integrate geophysical, ecological, social science, and economic data, especially investigations of the prediction and consequences of rapid change.*

Because of its broad interdisciplinary approach, research initiated in IPY 2007-2008 stands to make a significant contribution to our understanding of the causes and consequences of change in the polar regions.

Recommendation 2: The U.S. science community and agencies should pioneer new polar studies of coupled human-natural systems that are critical to U.S. societal, economic, and strategic interests.

• *Encourage research to understand the role of the polar regions in globally linked systems and the impacts of environmental change on society.*

Daily life and economic and strategic activities are constantly affected by changing environmental conditions, including the frequency and degree of severe weather events such as storms or droughts in many regions, including the continental United States. Investigations of impacts of linked environmental-technological-social change and health effects in many communities, including northern communities, are needed.

• *Investigate physical-chemical-biological interactions in natural systems in a global system context.*

Interdisciplinary approaches hold great promise for understanding the dynamics of anthropogenic activities, technologies, and environmental consequences. Investigations of linked atmospheric-oceanic-ice-land processes in the polar regions will enable understanding of global linkages and transformations due to natural and anthropogenic causes.

• *Examine the effects of polar environmental change on the human-built environment.*

Because of the recent large-scale environmental changes, northern communities, infrastructure, and other forms of human-built environment are affected by a variety of factors, such as the thawing of permafrost, higher frequency of severe storms and weather conditions; increased shore- and beach erosion, vegetation die-off, and fire danger. New engineering and policy research should investigate economically feasible and culturally appropriate mitigation techniques for countering the effects of a changing environment on technology, local communities, and their infrastructure, including all-season ground and air transportation, the design of roads, harbors, foundations, and buildings.

Recommendation 3: The U.S. International Polar Year effort should explore new scientific frontiers from the molecular to the planetary scale.

• *Conduct a range of activities such as multidisciplinary studies of terrestrial and aquatic biological communities; oceanographic processes, including seafloor environments; subglacial environments and unexplored subglacial lakes; the Earth's deep interior; and Sun-Earth connections.*

Opportunities for discoveries exist in many areas, and research could elucidate the structures of poorly understood biological communities, notably the microbial populations that contribute to most biogeochemical transformations; reveal oceanic processes that contribute importantly to biological productivity and climate; and discover new physical, chemical, and, potentially, biological characteristics of subglacial lakes long isolated from atmospheric contact. This research also could help understand major geological processes such as seafloor spreading, explore the subglacial topography and bedrock geology of regions important for Earth's climate history, map the structure of

Earth's interior and explore the links between mantle structure and surface processes, and provide an integrative synthesis of the interactions of our planet with the Sun.

- *Apply new knowledge gained from exploration to questions of societal importance.*

Polar biological studies, notably those that employ modern genomic methodologies, will advance biomedical and biotechnological research. For example, understanding how small mammals withstand temperatures near freezing during hibernation will contribute to improved protocols for cold storage of biological materials and for cryosurgery. Studies of oceanographic phenomena will facilitate more accurate understanding of the mechanisms driving climate change. Understanding how increased flow of fresh water into the polar oceans alters circulation patterns and transfer of heat from the tropics to the poles is one example of contributions from oceanography. Advances in the geosciences (e.g., through study of the extremely slow seafloor spreading rates in the Arctic) may shed light on tectonic processes that contribute to seismic events. Better understanding of solar influences on the atmosphere and Earth will improve understanding of the forces that drive weather systems and of solar activity on global communications and other technical systems.

- *Invest in new capabilities essential to support interdisciplinary exploration at the poles.*

New scientific discoveries are based in part on the availability of enhanced logistics to provide access to unexplored regions as well as new technologies to provide new types of data. The IPY field component should aggressively seek to further develop innovative strategies for polar exploration.

Recommendation 4: The International Polar Year should be used as an opportunity to design and implement multidisciplinary polar observing networks that will provide a long-term perspective.

- *Design and establish integrated multidisciplinary observing networks that employ new sensing technologies and data assimilation techniques to quantify spatial and temporal change in the polar regions.*

The IPY will provide the integrative basis for advancing system-scale long-term observational capabilities across disciplines. A goal of the IPY should be the design and establishment of a system of integrated multidisciplinary observing networks. New autonomous instrumentation requires development with the harsh polar environment in mind. Instruments required for different types of studies can be clustered together, minimizing the collective environmental risks of survival and encouraging integrated analysis. Common observational protocols, such as observation frequency and measurement precision, will increase the spatial range of the observations and simplify data assimilation. Once established in the IPY, such protocols will serve polar science in the longer term.

- *Conduct an internationally coordinated "snapshot" of the polar regions using all available satellite sensors.*

Two hallmarks of the IGY were the dawn of the satellite era and the establishment of enduring benchmark datasets. Today's ever-growing suite of satellite sensors provides unique views of the polar regions with unprecedented detail. Marshaling the collective satellite resources of all space agencies around the world would supply generations of future scientists an unparalleled view of the state of the polar regions during the IPY 2007-2008.

> **Recommendation 5: The United States should invest in critical infrastructure (both physical and human) and technology to guarantee that the International Polar Year 2007-2008 leaves enduring benefits for the nation and for the residents of northern regions.**

• *Ensure the long-term availability of assets necessary to support science in the polar regions, such as ice-capable ships, icebreakers, submarines, and manned and unmanned long-range aircraft.*

Although IPY 2007-2008 is planned as a focused burst of activity with demonstrable results, it should also provide long-term value and leave a legacy of infrastructure and technology that serves a wide range of scientific studies for decades to come.

• *Encourage development of innovative technologies to expand the suite of polar instruments and equipment, such as unmanned aerial vehicles (UAVs), autonomous underwater vehicles (AUVs), and rovers.*

Observational systems for the polar regions can be improved enormously by applying innovative technologies. Recent technological advances in UAVs, AUVs, and robotic rovers can be marshaled and adapted for the IPY to ensure that these platforms enhance IPY research capabilities.

• *Develop advanced communications systems with increased bandwidth and accessibility capable of operating in polar field conditions.*

The innovative technologies and large-scale field operations during IPY 2007-2008 will require advanced communications systems with high-speed, real-time access to communicate and distribute data from both polar regions to the rest of the world.

• *Develop international standards, policies, and procedures that ensure data are easily accessible for the current generation and permanently preserved for future generations.*

The data management systems should provide free and open access to data in standard formats. In addition, extensive metadata should be included to facilitate long-term reanalysis and so that datasets can be used by a variety of users. This effort should include data rescue efforts to expand the data record back in time and ensure that historical data are not lost.

• *Develop the next generation of scientists, engineers, and leaders and include underrepresented groups and minorities.*

Tomorrow's leaders are in today's classrooms, and the IPY effort should focus on cultivating an interest in the next generation of scientists, engineers, and leaders to create a lasting legacy.

Recommendation 6: The U.S. International Polar Year effort should excite and engage the public, with the goals of increasing understanding of the importance of polar regions in the global system and, at the same time, advancing general science literacy in the nation.

• *Develop programs in education and outreach that build on the inherent public interest of the polar regions and provide a broad lay audience with a deeper understanding of the polar regions.*

The polar regions have important direct and indirect effects on the rest of the world, and the IPY can help explain the importance of the polar regions to the public.

• *Create opportunities for education, training, and outreach for all age groups and build on successful existing models. Education and outreach during the IPY should include innovative new approaches that are interactive, make use of diverse media, and provide opportunities for hands-on participation by the public.*

The polar regions are inherently exotic to many people—the terrain, the plants, the animals, the weather, the remoteness—and they capture our imagination. This is key to engaging the public. There will be opportunities for formal classroom programs for people of a variety of ages, and media coverage that will provide both entertainment and enjoyable science education.

Recommendation 7: The U.S. science community and agencies should participate as leaders in International Polar Year 2007-2008.

• *Guide and contribute to IPY 2007-2008 activities and help to evolve the international framework, using the IPY as an opportunity to build long-lasting partnerships and cooperation across national borders.*

IPY 2007-2008 is an international effort, with more than 25 nations already committed to participate. Because of the strength of U.S. polar programs, our nation stands to play a leadership role in organizing and carrying out this ambitious program. Planning at the international level is under the auspices of two major organizations, the International Council for Science (ICSU) and the World Meteorological Organization (WMO), and the United States should lead the coordination with other countries through the ICSU and WMO to ensure the success of the IPY.

• *Continue to plan IPY 2007-2008 using an open, inclusive process.*

The initial impetus for organizing IPY 2007-2008 came from the science community, which has come together and worked diligently to identify activities of merit. This open process leverages the intellectual assets of the U.S. science community and should be continued.

• *Coordinate federal efforts to ensure a successful IPY effort, capitalizing on and supporting existing agency missions and creating new opportunities.*

International polar science efforts that have already been planned by the U.S. science community provide models for interagency collaboration, and additional future interagency efforts are encouraged, including coordination with the Arctic Council.

- *Continue planning for IPY 2007-2008, moving toward the creation of a more detailed science implementation plan.*

The next phase of IPY planning will need to provide concrete guidance that defines the science goals and addresses logistics and other key aspects of implementation. This phase of planning should include active participation by the U.S. science community and U.S. funding agencies and also continued efforts to coordinate with international planning activities so that resources are leveraged.

- *Provide mechanisms for individuals, early-career researchers, and small teams to contribute to the IPY.*

The overarching science goals of the IPY are broad and focused on international cooperation, but mechanisms for early-career researchers and small teams must be included in the larger IPY framework.

1

Why an International Polar Year in 2007-2008

"Whatever you can do, or dream that you can do, begin it.
Boldness has genius, power, and magic in it."

GOETHE

Environmental changes currently observed in the polar regions are unprecedented in times of modern observation, and there is concern that these rapid changes may continue or even amplify in the coming decades (IPCC, 1998, 2001). The harbingers of change can be seen vividly in the polar regions. The Arctic ice cover is decreasing in extent and area (Cavalieri et al., 1997; Johannessen et al., 2004); some ice shelves in Antarctica are retreating and thinning (Skvarca et al., 1999; Shepherd et al., 2003); glaciers across the globe are disappearing (Arendt et al., 2002); ecosystems are changing (Hunt et al., 2002); Alaskan villages, including Shishmaref, are being moved to higher ground in response to coastal erosion; and permafrost thawing is causing the collapse of roads and buildings (ACIA, 2004). Are we witnesses to the maximum in natural variability or the threshold of an abrupt change? How will changes first seen in the polar regions propagate and influence humans and the environment across Earth?

Events observed today in the polar regions represent a call to action to address many important broad and interlinked research challenges. Changes that we are witnessing in the polar regions today are unlike any in recorded history, yet we do not understand how or why the changes are occurring, and we lack the tools and knowledge to predict, mitigate, or adapt to the outcome. Changes in ice mass reflect that multidecadal integrations of small changes can lead to big changes; implementing polar observation systems is an essential step to document these changes. Clues for understanding how and why similar changes occurred in the past remain stored in polar earth and ice; analyses of sediment and ice cores are needed for understanding past changes. Polar changes are interlinked with the behavior and survival of ecosystems, from microbial life to large organisms, including humans; interdisciplinary polar studies in biology therefore are needed (NRC, 2003a). Keys to fundamental discoveries for understanding change may spring from new modes of exploration that range from using autonomous vehicles for ice studies to using genomics to investigate adaptation. Exploration reveals surprises.

The changes are not restricted to the physical environment; communications technologies, such as television and the internet, are challenging traditional human lifestyles in cold regions and elsewhere. Yet these same technologies hold the potential for essentially instantaneous sharing of information and for promoting global understanding. Internet-based efforts in global data collection, data sharing, and education hold tremendous potential.

In a world of much environmental uncertainty, citizens turn to science for answers. The polar regions are more sensitive to global climate changes, and therefore polar research plays an important role in providing answers (e.g., Albert, 2004; Stone and Vogel, 2004). To this end, scientists around the globe have come together to begin planning the International Polar Year 2007-2008. As described in the report by the International Council for Science's (ICSU) IPY Planning Group (Rapley and Bell, 2004), IPY 2007-2008 is envisioned as the dawn of a new era in polar science—it will be an intense, internationally coordinated campaign that gives expanded attention to the deep relevance of the polar regions to the health of our planet, and it serves to establish the ongoing observation systems, programs, and intellectual commitment

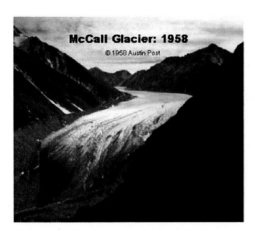

Photographs of the McCall Glacier in Alaska, located in what is now the Arctic National Wildlife Refuge, which has the longest history of scientific observation for any U.S. Arctic glacier. These observations began as part of the International Geophysical Year in 1957-1958 by Austin Post, who was given an honorary Ph.D. in May 2004 for his IGY contributions. The decrease in the ice extent is consistent with other glaciers across Alaska. Laser altimetry was used to estimate volume changes of 67 glaciers in Alaska from the mid-1950s to the mid-1990s and the average rate of thickness change of these glaciers was −0.52 cubic kilometers per year (water equivalent), equivalent to a rise in sea level of 0.14 millimeters per year. These recent losses of ice from Alaskan glaciers are nearly double the estimated annual loss from the entire Greenland ice sheet during the same time period and are higher than previously published loss estimates for Alaskan glaciers. They represent the largest glaciological contribution to rising sea level yet measured. SOURCES: Austin Post and Matt Nolan, University of Alaska, Fairbanks.

needed to fully understand the polar regions and their links to the global system. It will include research in both the Arctic and Antarctic, be multi- and interdisciplinary in scope, and be truly international in participation. It will educate and excite the public and help produce the next generation of engineers, scientists, and leaders. A framework such as the IPY can provide the impetus to undertake projects that could not be achieved by any single nation. It allows us to think beyond traditional borders—whether national borders or disciplinary constraints—toward a new level of integrated, cooperative international science.

NATIONS WORKING TOGETHER CAN ACCOMPLISH WHAT NO ONE NATION CAN DO ALONE

Nations around the world are making plans for IPY 2007-2008 to attempt to answer these and many more questions. Previous IPYs (1882-1883 and 1932-1933) and the International Geophysical Year (IGY; 1957-1958) produced unprecedented exploration and discoveries in many fields of research and fundamentally changed how science was conducted in the polar regions (see Box 1.1). IPY 2007-2008 will benefit society by exploring new frontiers and increasing our understanding of the key roles of the polar regions in globally linked systems. Recent technological developments give us a new ability to investigate previously unexplored areas, using new tools to understand

Scientists working in the polar regions, like these scientists traveling across the Arctic sea ice, face many challenges caused by the harsh conditions. IPY will bring scientists from many nations together to study some of the most important questions of our times. SOURCE: Jackie Grebmeier, University of Tennessee.

BOX 1.1
Past Polar Years and Their Contributions

On three occasions over the past 125 years scientists from around the world banded together to organize concentrated scientific and exploration programs in the polar regions. In each major thrust or "year," scientific knowledge and geographical exploration were advanced, thereby extending understanding of many geophysical phenomena that influence nature's global systems. Each polar year was a hallmark of international cooperation in science. The experience gained by scientists and governments in international cooperation set the stage for other international scientific collaborations. International scientific cooperation also paved the way for several political accords that gained their momentum from the polar years. IPY 2007-2008 will expand on this legacy of scientific achievement and societal benefits.

First International Polar Year (1882-1883). The fundamental concept of the first IPY was that geophysical phenomena could not be surveyed by one nation alone; rather, an undertaking of this magnitude would require a global effort. Twelve countries participated, and 15 expeditions to the polar regions were completed (13 to the Arctic and 2 to the Antarctic). For the United States it provided an opportunity to establish a scientific station at Point Barrow, the northernmost point in Alaska and the continental United States. Beyond the advances to science and geographical exploration, a principal legacy of the first IPY was setting a precedent for international science cooperation.

Second International Polar Year (1932-1933). The International Meteorological Organization (a predecessor to the World Meteorological Organization) proposed and promoted the second IPY (1932-1933) as an effort to investigate the global implications of the newly discovered jet stream. Some 40 nations participated in the second IPY, which heralded advances in meteorology, atmospheric sciences, geomagnetism, and the "mapping" of ionospheric phenomena that advanced radioscience and technology. Forty permanent observation stations were established in the Arctic, creating a step function expansion in ongoing scientific Arctic research. The U.S. contribution to this effort also included more stations in the Arctic. In Antarctica the U.S. contribution was the second Byrd Antarctic expedition, which established a winter-long meteorological station approximately 125 miles south of Little America Station on the Ross Ice Shelf at the southern end of Roosevelt Island. This was the first research station inland from Antarctica's coast.

The International Geophysical Year (1957-1958). The IGY (July 1, 1957 to December 31, 1958) celebrated the 75th and 25th anniversaries, respectively, of the First and Second IPYs and had participation from 67 nations. The IGY was conceived by a number of eminent World War II physicists, including Sydney Chapman, James Van Allen, and Lloyd Berkner, at an informal gathering in Washington, D.C., in 1950. These individuals realized the potential of the technology developed during the war (e.g., rockets, radar), and hoped to redirect the technology and scientific momentum toward advances in research, particularly in the upper atmosphere. The IGY's research, discoveries, and vast array of synoptic observations revised or "rewrote" many notions about the Earth's geophysics. IGY research helped solve the long-disputed theory on continental drift. During the IGY, the Earth's first artificial satellites were launched, and the Van Allen Radiation Belt encircling the Earth was discovered. For many disciplines the IGY led to an increased level of research that continues to the present. A notable political result founded on the IGY was ratification of the Antarctic Treaty in 1961. The success of the IGY also fostered an additional year of research through the International Geophysical Cooperation. The Special Committee for the IGY became the model on which three post-IGY scientific committees were developed, for Antarctic, oceanic, and space research, and several focused research efforts, including the International Year of the Quiet Sun. The scientific, institutional, and political legacies of the IGY endured for decades, many to the present day.

once-unanswerable questions. Automatic observatories, satellite-based remote sensing, autonomous vehicles, the internet, and genomics are just a few of the new approaches for studying previously inaccessible realms.

IPY 2007-2008 is being planned, from the start, as a truly international collaboration, with active planning from an international planning group and endorsements from a number of important international science organizations, including the ICSU and the World Meteorological Organization (WMO; see Box 1.2, Box 1.3, and Appendix A). This international framework is critical: IPY 2007-2008 will seek to marshal global scientific, human, and financial resources to address problems with a scope, complexity, and importance in ways that could not be accomplished by one nation alone. Development of the next IPY as a collaborative international effort has several tangible benefits, including:

1. *Wider participation.* Nations can participate under the umbrella of the IPY at a scale that their resources allow. For instance, larger nations such as the United States, which already plays a key role in polar research (NSB, 1987; NSF, 1997), can provide enhanced ground-, ice-, air-, and ocean-based stations and platforms, while smaller nations can provide specific instruments for a network of sensors. This partnered research also will lead to increased innovation. Countries seeking a common approach to a mutual problem benefit from the aggregated expertise. The technical means to accomplish the mutual goal are increased and the gain is shared more widely. The continued revolution in technology that is producing ever more capable components in smaller and lower power configurations is supplying new tools ideally suited for polar applications.

2. *Broader perspectives.* International collaboration provides multiple perspectives and capabilities. It offers the possibility of unique synergies of thought and optimizes shared resources, and it enhances our ability to understand one another as well as our world. Partnerships will be real and respectful, and will reflect the value of each contributor. This includes early engagement and full consultation during the initiation, implementation, and completion of all facets of these programs. Contributions of varying size and complexity, scaled to abilities and available resources, lead to the broadest participation.

3. *Increased data coverage and compatibility.* The involvement of a larger number of nations increases the potential spatial coverage of field programs. For instance,

---

BOX 1.2
Criteria for International Polar Year Initiatives

In its deliberations, the U.S. National Committee for the International Polar Year suggested the following criteria for IPY initiatives:

- Address compelling science issues in both polar regions
- Involve multi-national and interdisciplinary interactions
- Attract and develop the next generation of scientists, engineers, and leaders
- Engage the public

BOX 1.3
The International Planning Context

Discussions about holding an International Polar Year in 2007-2008 began simultaneously in many nations and many scientific settings. Initial discussions were informal, but this loose brainstorming turned quickly into more organized planning when the International Council for Science established an IPY Planning Group in the summer of 2003. Endorsement by the WMO in this same time frame gave the effort added weight. The ICSU IPY Planning Group, composed of leading polar scientists, and including members with links to other planned large-scale efforts being organized in celebration of the IGY anniversary, was instrumental in laying out the basic vision for IPY and calling on nations to take an active role in the continued definition of science themes, goals, and procedures. The ICSU IPY Planning Group's report to the ICSU Executive Board (Rapley and Bell, 2004) has been distributed and is posted online (http://us-ipy.org).

In an effort to understand what the science community saw as the most important opportunities for IPY 2007-2008, the planning group put out a general call for input. The response was positive and reassuring: more than 325 proposals were received from many individuals, institutions, and organizations. This enthusiastic response has invigorated the polar communities. Nations continue to express interest in participating, ideas continue to be submitted, and planners are working hard to design the next steps so that these general ideas can be organized into active programs and clear mechanisms are designed to allow wide participation. The next step is for planners, with input from all involved nations, to work with scientific experts and agency representatives to transform the many research activities into a coordinated set of cutting-edge research projects that fulfill the IPY vision—programs which are multi- and interdisciplinary in scope, truly international in participation, educate and inspire the public, and train the next generation of engineers, scientists, and leaders.

The U.S. research community is encouraged to participate fully in whatever planning activities occur within these international organizations to facilitate international cooperation at an early stage and to ensure that excellent research ideas are not lost. It is probable that some IPY research proposed by the U.S. National Committee will also be voiced in ideas proposed by other nations or by existing international scientific organizations. Common interests will be united by the ICSU IPY Planning Group.

The roster of potential IPY 2007-2008 nations is large and growing. In the North there are eight nations with territory and populations in the Arctic, all of which intend to participate (i.e., the United States, Canada, Russia, Greenland/Denmark, Iceland, Norway, Sweden, and Finland), but there are many more nations with research interests in the region, including Germany, Japan, and China who likely will participate in IPY 2007-2008. In the South, 45 nations are signatories to the Antarctic Treaty and thus have put aside territorial claims to share the continent for scientific purposes. There are also a number of international organizations with responsibilities related to polar research, such as the Scientific Committee on Antarctic Research, the Scientific Committee for Oceanic Research, the International Arctic Science Committee, the Arctic Ocean Studies Board, and others that are now actively involved. Another layer of involvement comes from existing international science programs that share interest in the science themes being developed, such as the World Climate Research Program's Climate and the Cryosphere project and the Scientific Committee on Solar-Terrestrial Physics' Climate and Weather in the Sun-Earth System program. Taken together, all of these different participants bring great momentum to IPY 2007-2008 planning.

nations can deploy similar systems, such as moorings, buoys, profilers, land- and ice-based systems, and share data in compatible formats. This international cooperation will result in benchmark and ongoing datasets collected by standardized methodologies. Differences in regionally based datasets remain a major impediment to global climate research. Temporally and globally consistent merged datasets from observing networks that span the full extent of the polar regions will provide calibrated base lines that are easily shared and that enable international research to flourish, ultimately allowing for scaling to larger spatial analyses of environmental change.

4. *Shared human experiences.* Less obvious, yet palpable, benefits of international collaboration extend beyond the scientific arena. Apart from the advantages gained by humans when the climate of our planet is better understood, there is shared human experience in exploration that is deepened the more challenging the environment and the more hard-won the understanding. The IGY resulted in a number of unanticipated geopolitical accomplishments despite the tense political climate of the day. IPY 2007-2008 will incorporate this human experience not only by bringing together scientists into international teams but by allowing people across the planet to share in this experience.

PURPOSE AND FRAMEWORK OF THIS REPORT

The International Polar Year 2007-2008 will initiate a new era of sustained, internationally coordinated polar science. This report reflects the vision of the U.S. Committee for the International Polar Year for participation in that new era. The purpose of this report is to present an overview of potential science themes, enabling technologies, and public outreach opportunities submitted by individuals and the science community in the United States for the next IPY. The objectives are to build consensus on the possible content of the IPY and ensure that plans for its activities address the key IPY goals of addressing issues relevant to both the Arctic and the Antarctic, involving scientists from a range of disciplines and nations and focusing on compelling scientific questions. But this report is also written to engage people other than scientists in thinking about IPY 2007-2008. The U.S. Congress recently showed interest in celebrating the 50th anniversary of the IGY, and this report provides strong evidence that the U.S. and international science communities are already engaged in planning extraordinary activities. This report presents the vision that articulates the overarching science from ideas voiced by many U.S. scientists. This report is not a science plan, for that effort will unfold in the coming months as communities and agencies work together to further define specific IPY activities and to determine exactly what will occur during the 24 months of the IPY and lay the groundwork for what will occur in the following years.

Information for this report was generated by a number of routes, including community information gathered through an online discussion forum, public comments on an IPY White Paper, solicited written summaries of research project ideas, committee expertise, and town hall meetings and special sessions at a variety of scientific conferences. Research ideas contributed by individuals and groups to the U.S. National Committee for the International Polar Year were insightful and wide ranging, and after deliberation the committee concluded that there were five scientific challenges (Chapter 2) and two very broad science themes: understanding change in the polar regions (Chapter 3) and exploring new frontiers (Chapter 4). These two overarching themes

This Antarctic sunset near the McMurdo base shows the "calving" of icebergs from the floating Ross Ice Shelf—the largest ice shelf in the world. Larger than the State of Texas, it measures about 660 feet (200 m) thick at the outer edge and about 2,300 feet (700 meters) thick along the inner edge. SOURCE: George Somero, Stanford University.

lay the foundation around which IPY activities can be organized. They are by definition very broad, to allow scientists to begin developing more specific science objectives during the next phase of planning.

In many cases, innovative technologies and techniques will be needed to answer scientific questions. Chapter 5 outlines some of the enabling technologies that hold special potential in achieving the objectives of the next IPY. Finally, the committee and the community were adamant that public outreach and education be fundamental for the next IPY, and Chapter 6 outlines the key characteristics needed in such activities and provides examples to lead potential participants to think about how to build outreach into all aspects of IPY 2007-2008.

In summary, this report is intended to present a first vision for U.S. participation in IPY 2007-2008. As international planning continues, research agency leaders continue discussions, and specific activities are developed, the vision for the IPY will sharpen. The next step in IPY planning will be to move from this broad vision, both within each participating nation and at the international level, toward more concrete activities. The mechanisms for this next step are still taking shape. The U.S. National Committee for the International Polar Year will continue to serve as a focal point for interaction with the international community and will seek ways to move the IPY from conceptualization to implementation.

Research at the Bottom of the World

Diana Wall, professor and director of the Natural Resource Ecology Laboratory at Colorado State University, is in love. More precisely, she is enchanted by invertebrates in the soil—"those under-appreciated animals that frame the web of life." Wall's research addresses the importance of soil biodiversity for ecosystems and society.

Working in the Antarctic is "crazy," she freely admits, and yet she can't stay away. She is in awe of "the vastness, and the fact that I am on a landscape that is geologic history in front of my eyes—much like Mars."

And it's important. "Understanding the polar regions is to global ecosystems what understanding model organisms is to human health," she says. "They are inseparable." The risk of not studying the polar regions is inconceivable, she adds. There are so many questions to answer—questions about weather, dispersal of organisms, evolutionary biology, and much more.

Specifically, Wall investigates how soil biodiversity contributes to healthy productive soils. She also is concerned with the consequences of human activities on soil sustainability. Fueled by this passion, Wall has spent 13 field seasons since 1989 in the cold desert ecosystem of the Antarctic Dry Valleys, which she says is "thrilling." Here she is learning how glaciers, soils, streams, and lakes are interconnected, and how biota respond.

In many ways, the Dry Valleys are among the harshest and most isolated environments on the planet, says Wall. The average annual temperature ranges between a chilly −4°F (−20°C) and 3.2°F (−16°C). In winter a cloak of total darkness descends upon the valley for a long time, and temperatures may plummet to −49°F (−45°C). At that time, each ecosystem adopts its own extravagant survival strategies as ordinary linkages between ecosystems are severed.

Because of these extremes in temperature and dryness, few species of plants and animals make their home in the Dry Valleys, Wall explains. Eukaryotic algae, cyanobacteria, and mosses comprise the stream communities, and microbial communities inhabit the glaciers and lakes. Food chains appear to be unusually short. So short, in fact, that the major polar "predator" in the valleys is the nematode—which is microscopic. The dominant invertebrate in this particular ecosystem, nematodes are present in 60 to 80 percent of the soils in the Dry Valleys, where they consume bacteria, fungi, and other microscopic animals.

Working in the Antarctic has taught Wall much about planning and patience, she says. This is because you decide which pieces of equipment and gear you need—from Petri dishes to tents—to have a successful trip. And you design the experiments.

"You plan this months in advance," she says. "And then, when you finally get to the ice, you have to be patient. Ultimately, the weather controls whether you have a good field season and get the experiments set up.

"If it's great weather, then I may get data from all the experiments I planned," says Wall. "If not, we have to make some hard choices about what are our priorities.

"It's crazy," she says again. "It is hard, long, exhausting, and I love it."

Photo of Diana Wall. SOURCE: Emma Broos.

2

Scientific Challenges for the International Polar Year

"Results of much greater scientific value can be expected if standard observations are made by observers using similar instructions for recording phenomena at simultaneous periods, and who exchange the results of their observations without discrimination."

KARL WEYPRECHT 1879

The polar regions play a critical role in linked atmospheric, oceanic, terrestrial, and biological processes occurring on local and global scales. Changes first seen in the polar regions affect weather and life in many areas, including the mid-latitudes. The International Polar Year 2007-2008 is an opportunity to deepen our understanding of the physical, biological, and chemical processes in the polar regions and their global linkages and impacts and to communicate these new insights to the public at large. The most logical strategy is an international, coordinated, and intense scientific program of polar study. Such a program would not only add to our understanding of environmental change and scientific exploration, it would result in a world community with shared ownership in the results. Five broad scientific challenges provide a framework for organizing IPY 2007-2008 activities:

ASSESSING LARGE-SCALE ENVIRONMENTAL CHANGE IN THE POLAR REGIONS

Environmental observations demonstrate that dramatic change on timescales from years to many millions of years is a hallmark of Earth's history. Comparatively small changes in climate in the past 10,000 years have had a profound effect on humans, while comparatively larger abrupt climatic changes have occurred repeatedly in the past 100,000 years (NRC, 2002). Large-scale environmental changes in the polar regions within the past few decades are more pronounced than changes in the mid-latitudes or tropics, and many of the changes being witnessed involve poorly understood linked regional and global processes. Many scientists are concerned that the climate system is being pushed toward a warmer state that likely will have temperatures higher than at any time during Earth's current glacial/interglacial cycle. Yet we have an insufficient scientific basis on which to determine whether the changes are due mainly to anthropogenic effects or the effects of natural variability. Whether the changes are attributable to anthropogenic or natural causes, the impacts will still affect humans and the environment. Therefore, we must understand present changes in the context of past changes in order to develop a basis for societal response to changes occurring on our small world.

SCIENTIFIC EXPLORATION OF THE POLAR REGIONS

In many areas the changes and their causes are only partly perceived because the polar regions are not completely "mapped," and exploration of such elements as the seafloor, the ice sheet bed, the crustal domain, and the biota is still needed to understand fully the nature and cause of past changes. While exploration will lead to answers for important geological, climatological, glaciological, and biological questions, it also will provide unexpected discoveries. The polar regions represent a biological extreme inhabited by a rich diversity of organisms uniquely adapted to their environment. Understanding this diversity, including how it functions and responds to its environment, will lead to discoveries of scientific, medical, and industrial importance (NRC, 2003a). New logistical capabilities and recently developed technologies such as autonomous vehicles, high-speed communications, and high-throughput genomic tools, coupled with international collaboration, will enable these regions to be studied in a fundamentally new way that was not possible even several years ago.

OBSERVING THE POLAR REGIONS

Our ability to describe the environmental state of the polar regions, including variability, change, and extremes, is critically limited by a lack of observations in space and time. Records of past environmental conditions, retrieved from paleoarchives such as ice cores or sediments, provide clues to nature's response to forcing, but these too are incomplete, especially in terms of spatial coverage. Recent technological advances, including satellite observations and remotely operated equipment, demonstrate that major breakthroughs in scientific understanding of the extreme environments are possible. Realization of this potential for new insights poses a major challenge in the development of innovative technologies, the design of system-scale observing systems, and the development of strategies for interdisciplinary observations. An internationally coordinated approach will allow us to greatly advance our knowledge of the polar regions through enhanced capability to observe it with the required spatial and temporal resolution.

UNDERSTANDING HUMAN-ENVIRONMENT DYNAMICS

The ways that wildlife and humans influence and are influenced by the polar regions are parts of the picture of environmental change everywhere but occur more vigorously in the polar regions, making the high latitudes a natural laboratory for gaining a better understanding of the consequences of change to humans and to wildlife. In today's world, society, technology, and the environment are closely linked. Because of atmospheric circulation patterns, large-scale environmental changes may result in regional impacts that differ from the global scale, for example, thawing of permafrost in some regions, including Alaska. This affects human daily life and economic and strategic activities, including communications technologies; ground, air, and sea transportation; and failure of roadways and foundations under freeze/thaw conditions. Investigations of linked environmental-technological-social changes on human health in northern communities could enable new understanding that is applicable elsewhere. Investigations of linked physical, chemical, and biological transformations in the polar regions will enable understanding of natural reaction and contaminant impacts on the environment. Interdisciplinary science holds great promise for understanding the strong links between rapid changes in the environment, technology, and the actions of individuals and societies. The next IPY offers many venues to study these issues internationally and to develop methods for resilience.

CREATING NEW CONNECTIONS BETWEEN SCIENCE AND THE PUBLIC

Important new findings from polar research require effective communication. If the IPY results are to be long-lasting, the public must be significantly engaged in the long term and provided with an ever-increasing understanding of the global importance of the polar regions. Sparking the public's interest in these regions will not be difficult. Americans of all ages are fascinated by the polar regions, the history of their exploration, their cultures, and the scientific research that results. In addition, human activities are affecting the polar regions; increasing public awareness that the polar regions play a key role in the Earth system, and may be harbingers for detecting some environmental change, is of critical importance. However, effectively communicating our scientific knowledge to the public is a significant challenge. Innovative use of the media and the internet will provide real opportunities for the public to experience and become involved in the IPY. The involvement of teachers, students, and scientists working together will engage current students to consider science as a career. These efforts will expand the public's understanding of science in general and promote international understanding.

Research at the Top of the World

Environmental chemist Mark Hermanson is a curious guy. Since the 1970s, scientists had a hunch that elevated amounts of human-generated toxins might exist in such areas as the Arctic, thousands of miles from their industrial or agricultural sources. None of these early studies measured the actual inputs of pollutants, however, so their results were uncertain. Hermanson waited for some conclusive evidence, but none came.

Finally, he could wait no longer. He packed up his gear and flew to a remote area of the Hudson Bay to get some answers. There he did what no one else had ever done. He swam to the bottom of the Arctic lakes and collected sediment cores to bring back to the States to analyze. The cores confirmed the presence of elevated amounts of lead, cadmium, copper, mercury, and zinc—all potentially toxic heavy metals—that could only have been carried to the Arctic through the atmosphere.

As a professor of earth and environmental science at the University of Pennsylvania, Hermanson tracks the transport, fate, and effect of pollutants in atmospheric pathways around the Earth. While his most recent work is concentrated on the glacial islands of Svalbard, in Arctic Norway, much of his overall research has centered on the Inuit Hamlet of Sanikiluaq, on the Belcher Islands in Arctic Canada. "Most of the metals we have observed in sediments collected around the Belcher Islands are not yet indicative of a problem and appear to be declining, in fact. Still, our most recent data suggest that additional mercury continues to move into the area from the atmosphere."

Hermanson is concerned because the majority of the Inuit diet consists of the fish and seal the people hunt from the waters around their islands. They will eat birds and bears when they can get them he says, but there is a restriction on the number of bears that may be hunted, so they eat fewer of these than in the past. Mostly they eat fish, and fish can concentrate high levels of mercury in their tissues.

Widespread contamination of a major source of protein would be problematic to any community. It may be more of a concern in an Inuit community, Hermanson says, in part because their food alter-

Mark Hermanson swims to the bottom of Arctic lakes to collect sediment cores to monitor toxins carried into the Arctic through the atmosphere. SOURCE: Jeremy Stewart.

natives are so limited. Because of the great distances involved, for example, fresh fruits and vegetables are a luxury for the most part. In addition, because the Inuit are hunters by tradition, as were their ancestors for a thousand years before them, they have a rich repertoire of cultural rituals and beautiful legends about their way of life that they pass on to their children.

Hermanson has been traveling to the Arctic for more than 20 years. The work is important but difficult, too, he says. It takes a lot of energy, money, and many weeks of preparation. "It can be gruesome at times, too" he adds, remembering that once he was nearly lost in a white-out that came on very quickly. "But then I think of the vast treeless terrain, and the aurora dancing across the sky. And I haven't met any polar bears yet!"

3

Understanding Change in the Polar Regions

"When we try to pick out anything by itself,
we find it is tied to everything else in the universe."

JOHN MUIR

From the collapse of some fisheries to the disintegration of sea ice and floating ice shelves, recent change in the polar regions has captured the attention of the American public. Scientists predict the environment will continue to change in response to increasing levels of greenhouse gases in the Earth's atmosphere, with particularly high sensitivity in the polar regions (IPCC, 1998, 2001). The primary scientific issues are (1) to understand the nature of changes we are experiencing today within the context of the past, in order to discriminate between anthropogenic and natural variability and (2) to understand global linkages and the coordinated impact of changes on climate and weather patterns, ecosystems, and human endeavors across the planet. The Earth has strongly linked systems, and it is important to understand mechanisms and impacts of changes in the polar systems. Rapid changes, in particular, can be devastating if they exceed the timescale of adaptability, and there is compelling evidence from past climate records that the environment is capable of changing abruptly, on timescales of ten years or less (NRC, 2002).

HUMAN-ENVIRONMENT DYNAMICS

Studies of a wide range of issues related to the human dimensions of change in the polar regions stand to make important International Polar Year contributions. These human-environmental linkages directly involve the people who live in northern regions in many ways, but also involve people all around the world. For example, at this global level, changing environmental conditions—including the frequency and duration of severe weather events such as storms, precipitation, or drought—can have a range of impacts on human daily activity in temperate as well as polar regions. The Arctic Oscillation/North Atlantic

Oscillation, the El Niño-Southern Oscillation, and the Antarctic Oscillation are all multiyear low-frequency patterns of atmospheric and oceanic circulation that have effects ranging from major flooding in some regions to droughts and fires in others (NRC, 2002).

Changing environmental conditions affect systems we use every day, from the impact of snow on the mobility of vehicles and freeze/thaw destruction of roadways to icing of aircraft wings, and these pose engineering challenges. Permafrost has received much attention recently because surface temperatures are rising in most permafrost areas of the Earth, bringing some permafrost to the edge of widespread thawing and degradation. In the Arctic, thawing permafrost due to warming is resulting in the loss of soil strength. This has already caused the failure of roadways, runways, and pipelines

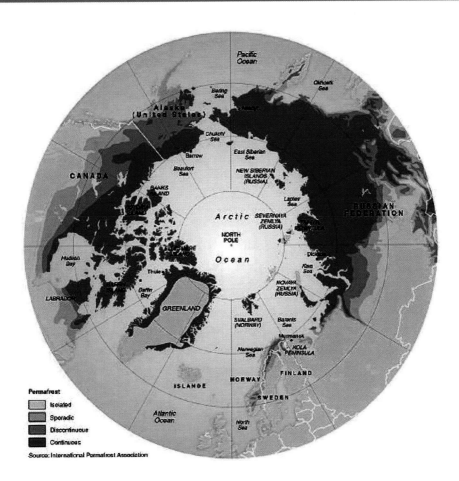

Permafrost, found mainly in the polar regions, is ground that has a temperature lower than 0°C continuously for at least two consecutive years. In the continuous zone, permafrost is found almost everywhere, while in the discontinuous zone, permafrost is found intermittently. In the sporadic zone, permafrost is found in isolated, small masses, and in the isolated zone permafrost is infrequently found. SOURCES: International Permafrost Association and UNEP Global Resource Information Database.

and is causing the foundations of some structures to collapse (ACIA, 2004). These observed and predicted changes in permafrost make it important to monitor permafrost dynamics (particularly its temperature through boreholes) so that we become more skilled at assessing and projecting possible impacts on ecosystems and infrastructure.

Another possible human-environment interaction linked to environmental change relates to changing ice conditions in the Arctic. Continued loss of summer sea ice could increase ship access in the Northwest Passage along the northern shores of Canada and the Northern Sea Route (Northeast Passage) along the Arctic coast of Russia (INSROP, 1999), with implications for commerce and national security (ONR, 2001). Due to large-scale atmospheric flow patterns, continued global warming may result in cooling in some north Atlantic regions (Alley, 2000).

One clear illustration of a polar-global linkage with a direct human impact is the transport of contaminants from industrial regions to the Arctic. The discovery of "Arctic haze" in the 1970s and early 1980s (Barrie, 1986) demonstrated that the Arctic is not a pristine environment isolated from human activity elsewhere but rather a region well connected to natural and anthropogenic sources of chemicals by winds, ice movement, and marine currents. The study of this phenomenon led serendipitously to the discovery of ozone depletion in the troposphere in the Arctic marine boundary layer at polar sunrise (Oltmans, 1981; Bottenheim et al., 1986). It has only recently been realized that natural reactions in snow can have important effects on the troposphere (Dominé and Shepson, 2002). For example, ozone is perturbing the biogeochemical cycle of mercury, and tropospheric ozone depletion chemistry is likely to have a significant impact on near-surface radiative transfer, affecting the air-snow-sea exchange of biologically mediated compounds (Lu et al., 2001; Shepson et al., 2003).

The transport of contaminants such as heavy metals, pesticides, and persistent organic pollutants to the north has direct impacts on residents. Fish caught in remote northern lakes and oceans contain unexpectedly large amounts of mercury (AMAP, 2002). Observations in the Arctic showed evidence of anthropogenic pollution as early as the 1940s. Figure 3.1, from McConnell et al. (2002), shows the record of atmospheric pollution by lead as recorded in a Greenland ice core. A clear increase in lead is seen during the Industrial Revolution. This is followed by a decrease toward pre-anthropogenic levels when the use of unleaded gasoline became widespread and air quality laws were put into effect. Human activities do impact the environment on a large scale, and through wise policy decisions, improvements are possible. Future changes in climate could lead to spatial and temporal changes in transport patterns and deposition of pollutants. Complex interactions between the atmosphere, oceans, biota, and snow cause transformations that are not understood. International collaboration will promote understanding of linked physical-chemical-biological interactions occurring in the atmospheres, oceans, permafrost, snow and ice, and human food chains in the polar regions.

Beyond these physical impacts of the changing environment on people, there is strong concern about how rapid social and economic changes continue to affect indigenous cultures in the northern high latitudes. Some of these impacts are undoubtedly positive: satellite communications now link even the most remote northern communities to the rest of the world, allowing, for example, doctors to render care using telemedical technologies and northern hunters to travel with increased safety using global positioning systems as navigation aids. In addition to the many benefits of advances in technology and communications, rapid cultural change has also been

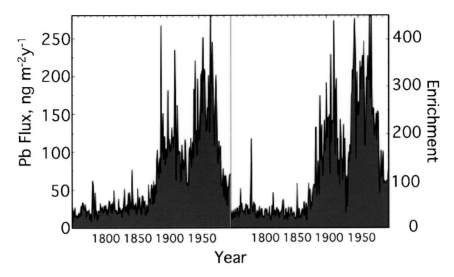

FIGURE 3-1 Continuous, high-resolution lead flux and crustal enrichment since 1750 from a 1999 ice core from Summit Greenland. Trends of increasing levels during the Industrial Revolution were reversed after air quality laws were put into effect. SOURCE: McConnell et al., 2002.

accompanied by problems such as alcoholism, drug use, and other social disorders in some northern communities, as it has in lower latitudes. There are concerns that environmental or social change could lead to problems with food safety and availability and circumpolar health in general. Understanding societal changes in the polar regions can provide lessons relevant to the broader science community as well as to the residents of lower latitudes who also face the impacts of social and environmental changes. More research is needed to better understand how new technologies can help northern indigenous peoples preserve their own cultural heritage, rather than allowing the same technologies to simply play a homogenizing role.

The firsthand impacts of environmental change are being felt by Arctic subsistence communities, such as those who hunt whales on the Alaskan North Slope, practice reindeer husbandry in Finland and Russia, and hunt caribou in northern Canada and Alaska (Krupnik and Jolly, 2001; Putkonen and Roe, 2003). Coastal erosion caused by unusual and irregular storm patterns and rising sea level have threatened coastal communities in the Arctic. U.S. government agencies have taken emergency actions to cope with the increased beach erosion and relocation of some communities in northern Alaska due to sea-level rise. Northern communities in permafrost regions will face significant engineering and infrastructure challenges if the permafrost thaws.

Modern technologies also can be affected by changes in the environment. The solar processes that produce disturbances in the Earth's space environment (space weather) affect high-frequency communications, including cell phones, global positioning systems, and power systems. Changes in ocean circulation and temperature patterns have an impact on acoustic propagation pathways for subsea communications. Changes in the patterns and severity of winters on land affect many modern

NASA satellite imagery analyzed at the University of Colorado's National Snow and Ice Data Center revealed that a section of the Larsen B ice shelf on the eastern side of the Antarctic Peninsula shattered and separated from the continent in early 2002. A total of about 3,250 square kilometers disintegrated in a 35-day period, an area larger than the state of Rhode Island (2,717 square kilometers). This is the largest single event in a series of retreats by ice shelves along the peninsula over the last 30 years. SOURCE: NASA.

technologies, from snow and ice impacts on ground travel to ice formation on aircraft wings (NRC, 2004a).

Interdisciplinary science holds great promise for understanding the strong links between rapid changes in environment, technology, and the actions of individuals and societies. Environmental, technological, and societal changes in the polar regions are occurring now, and understanding those changes can provide lessons relevant to all nations. The IPY offers many venues to study these issues internationally and to develop methods for resilience.

CHANGES IN THE POLAR REGIONS IN RECENT TIMES

Large-Scale Environmental Change

The character of recent changes in the polar regions (Table 3.1) results from the unique environmental conditions. Both polar regions have vast relatively permanent ice sheets that serve as massive freshwater reservoirs. In contrast, sea ice is typically only a few meters thick and is highly dynamic from year to year. Seasonal and interannual variations of sea ice can alter surface air temperatures in excess of 20°C locally, with far-reaching consequences. Surface warming melts ice and snow-covered surfaces, which normally are highly reflective, increasing the absorption of sunlight and, in turn, increasing the warming and melting. Satellite data and surface-based

TABLE 3.1 Evidence of recent environmental change in the polar regions

Parameter	Recent Evidence of Change
Surface air temperature	Significant temperature increases have occurred on the Arctic and the Antarctic peninsula.
Sea ice cover and ice sheet mass balance	The Arctic sea ice and snow cover has diminished considerably over the past two decades, particularly in summer, consistent with indigenous observations of early sea ice break-up and later freezes. Winter ice and snow conditions are also less stable. In addition, most glacial ice sheets in both polar regions have thinned, millennia-aged floating ice shelves are disintegrating, iceberg calving in the Antarctic has increased in frequency, and rapid rates of ice sheet thinning have been observed.
Ocean circulation	Dramatic changes in the circulation of the Arctic Ocean include a) the relocation of the boundary between the Atlantic and Pacific domains from the Lomonosov Ridge to the Alpha Mendeleyev Ridge system, and b) warming of the Atlantic layer and erosion of the cold halocline over the Eurasian Basin. The latter phenomenon exposes the sea ice cover to increased heat from below.
Permafrost	Permafrost degradation or thawing is occurring in northern Alaska, Canada, Russia, Mongolia, and China.
Pollution	Episodes of smog-like Arctic haze plague the spring surface layer, and high levels of toxic chemicals (Persistent Organic Pollutants, heavy metals, radionuclides) have been found in the Arctic environment.
Fisheries	The Bering Sea fisheries have been changing for decades, with a rapid decline in salmon and crab populations observed in the 1990s. There also has been a possible increase in shrimp population near Kodiak Island.
Birds and flora	Significant modifications to the distribution and migration patterns of many terrestrial and marine animal and plant species have been observed. These data are confirmed and substantially expanded by the observations of local residents, who report major changes in wildlife distributions, earlier spring arrival of many migrating animal and bird species, and the sighting of previously unknown life forms in the polar regions. There also are significant increases in greenness in the Arctic, particularly the expansion of shrubs.
Water cycle	Research indicates enhanced precipitation over the Antarctic peninsula and increased river runoff in northern Eurasia.
Carbon cycle	A decrease in sea ice extent will decrease ice algal production with impacts on higher trophic levels. Earlier sea ice melt in the Bering Sea has coincided with declines in carbon flux to sediments, faunal populations, and higher trophic levels.

SOURCES: Serreze et al., 2000; IPCC, 1998, 2001.

observations indicate that Arctic sea ice areal coverage has declined about 7 percent since 1978 (Johannessen et al., 2004). Change is sometimes dramatic: several times in the past decade, Arctic and Antarctic ice shelves, which are the extension of ice sheets into the ocean, have undergone spectacular breakups (Scambos et al., 2000; Mueller et al., 2003).

Even more significant is the predicted complete loss of the multiyear sea ice by the year 2100. The consequences of ice loss range from increased coastal erosion to an altered Arctic heat budget and increased safety risks on native coastal ice hunting grounds. The loss of summer sea ice not only endangers ice-endemic species like polar bears and the food web producing their prey but also will impact the biological cycles of the entire Arctic from the coast to the deep sea (Gradinger, 1995). As the sea ice disappears, understanding its role in the Arctic's biological cycles is an urgent issue critical to predicting the consequences of changes for the entire Arctic.

The climate of the Arctic has undergone rapid and dramatic shifts in the past, and there is no reason it could not experience similar changes again in the future. Records such as ice cores and sediment show climatic cycles that have occurred regularly on timescales from decades to centuries and longer and that are most likely caused by oceanic and atmospheric variability and variations in solar intensity. The Little Ice Age and Medieval Warm Period were examples of long-term cooler and warmer climates, respectively, while shorter-term decadal cycles such as the North Atlantic Oscillation and the Pacific Decadal Oscillation, among others, have been found to affect the Arctic climate. Since the industrial revolution in the nineteenth century, anthropogenic greenhouse gases have added another major climate driver. In the 1940s the Arctic experienced a warm period, like much of the planet, although it did not reach a level as warm as that evident in the late 1990s (Figure 3.2). It is the broadly accepted consensus of the science community that most of the global warming observed in the past 50 years is attributable to human activities (IPCC, 2001), and there is new and strong evidence that in the Arctic much of the observed warming over the same period is also due to human activities (ACIA, 2004).

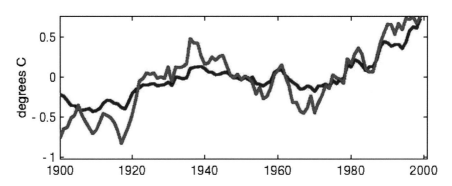

FIGURE 3-2 Records of annual mean surface air temperature for the Northern Hemisphere (blue) and the Arctic (red) from land-station data show that air temperatures above land surfaces in the Arctic appear to have warmed at a faster rate than the whole Northern Hemisphere during the twentieth century. The Arctic record shows a great deal of internal variability. Warming in the early part of the century is exceptionally large in the Arctic, as is the cooling that follows, which could indicate that this phenomenon originated in the Arctic. Alternatively, the enhanced response in the Arctic could be due to its sensitivity to external forcing. SOURCE: Jones and Moberg, 2003.

The world's oceans transport an enormous amount of heat poleward, moderating the global temperature. The polar oceans play a key role in forming the cold dense waters that sink in polar regions and drive ocean heat transport (Johannessen et al., 1994). One hypothesis to explain the suppression of recent change in the Antarctic relative to the Arctic is that ocean circulation changes around Antarctica have greatly increased the uptake of heat from the atmosphere in the Southern Ocean. Global climate models support this hypothesis and also predict that change in the Antarctic will eventually catch up with the Arctic near the end of this century (Manabe and Stouffer, 1994).

The polar atmosphere is another interactive link with the rest of the globe. Of the three global-scale fundamental modes of climate variability, two are rooted in the polar regions, the northern hemispheric annular mode and the southern hemispheric annular mode. The polar regions are at the center of hemispheric-scale circulation patterns, which coordinate variability in the environment across continents and ocean basins. These prominent circulation patterns have exhibited trends in recent decades that have a clear imprint on the spatial pattern of changes in air temperature, sea ice, snow cover, and storms, as well as many other components of the environment (e.g., Thompson and Solomon, 2002).

The land surface in the high northern latitudes is characterized by expansive snow cover in winter, followed by sharply peaked runoff and intense vegetation growth. Snowfall and rain have increased in many areas in recent decades, and so has discharge from many Arctic rivers (Serreze et al., 2000). An increase in rain in winter creates ice layers that are difficult for caribou and reindeer to paw through and that have been found to reduce their populations. Warmer and moister conditions are known to alter tundra vegetation, which in turn influences carbon uptake in the vast northern lands. Changes in snow cover also affect human activity. Exploration for oil and gas reserves on the North Slope in Alaska are only allowed when snow cover is at a depth adequate to protect the underlying tundra ecosystem, and the window of deep-enough snow cover has been decreasing in recent years (NRC, 2004a).

Ecosystems

Arctic and Antarctic systems are rich and diverse habitats for life. Key biogeochemical cycling processes occur in this extreme environment, being directly influenced by sea ice and snow cover, seawater hydrography (nutrients, salinity, and temperature), variable light levels, and atmospheric conditions. Polar organisms have adapted over time to live in an extremely cold environment and thus are among the most intimately susceptible of all species to climate warming events. Rising temperatures threaten the structures of their habitats as well as the function of their physiological processes.

Arctic sea ice is melting, and associated reductions in the extent of ice cover and ice thickness, and the impacts of these changes on the biological system, are dramatic and potentially devastating to certain species (Krajick, 2001; Whitfield, 2003). If the sea ice summertime coverage continues to decline (Figure 3.3), marine ice algae also would decline due to loss of substrate. The supply of algae has a cascading effect to higher trophic levels in the food web: small planktonic animals (zooplankton) feed on algae; many fish, such as cod, feed on these zooplankton, and sea birds and mammals in turn feed on the fish (Eastman, 1993). Species like polar bears would also be affected by a loss of sea ice as they almost entirely depend on it for foraging, hunting, and migrating.

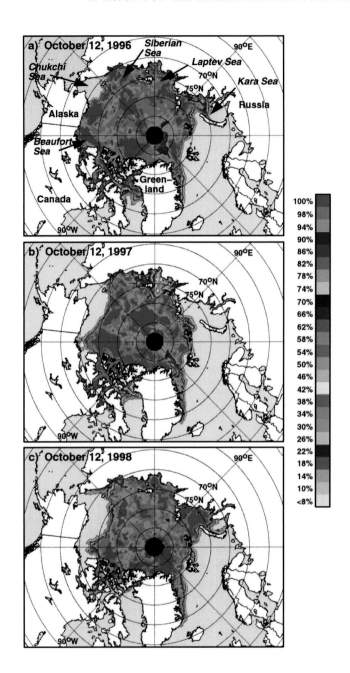

FIGURE 3-3 Reduction in sea ice extent in the Beaufort and Chukchi Seas over a three-year period. The ice cover in 1998 was a then-record-minimum low north of Alaska for the satellite era (post-1979). Since 1998, a new record low was observed in the summer of 2002 (Serreze et al., 2003), and 2003 was a close second (Fetterer et al., 2004). Sea ice concentration estimates derived from passive microwave satellite data. SOURCE: Comiso et al., 2003.

Thinning Ice Threatens Polar Bear Habitat and Polar Food Web

The primary natural habitat of the polar bear is under increasing threat as a consequence of thinning of the Arctic sea ice, which has decreased between 10 and 40 percent since the 1960s. Polar bears rely on the ice to hunt for seals, and its earlier breakup is giving them less time to hunt.

Looking ahead, some climate models predict a total loss of multiyear sea ice in the north by 2100 (USGCRP, 2001). What exactly would this mean for endemic species like the polar bear and the food web that produces their prey? Microscopic Arctic ice algae, for example, contribute as much as 50 percent of the total primary production at the bottom of the food web in high Arctic basins. The loss of the ice would have rippling consequences through the entire ecosystem.

SOURCE: Mark Weber, U.S. Fish and Wildlife Service.

Another indicator of contemporary Arctic change is the occurrence of an intense phytoplankton (coccolithophorid) bloom that is normally associated with warmer temperate regions. In 1997 such a bloom caused a massive die-off of short-tailed shearwaters, a seabird that annually migrates from nesting grounds in Australia to forage in the Bering Sea of Alaska (Hunt et al., 2002). Coincident with these changes was a buildup and then a crash in the biomass of large jellyfish (Brodeur et al., 2002). Other changes include declines in bottom-dwelling clam populations in the shallow northern Bering and Chukchi shelves in the 1990s, with these prey tightly linked to marine mammals and birds that are consumed by northern residents. Migrating gray whales shifted their feeding areas farther north in the 1990s, coincident with declines in bottom-dwelling shrimp-like amphipods (Moore et al., 2003). Studies in the southeastern Bering Sea suggest a reduction in overall productivity in the region (Schell,

Although primarily a high latitudes phenomenon, the northern lights have been seen as far south as Texas and are regularly seen at latitudes equivalent to New York. While historical references date as far back as centuries before Aristotle, it was not until the twentieth century that these celestial lights were understood. The dancing colors of the aurora reflect a large-scale electrical discharge phenomenon associated with oxygen and nitrogen atoms. SOURCE: Jan Curtis, University of Wyoming.

2000), whereas benthic (deep) population declines in the northern Bering Sea during the 1990s are coincident with reduced transport and a freshening of waters transiting the Bering Strait (Grebmeier and Dunton, 2000). Similarly, some studies have hypothesized and others have shown that fisheries are displaced as they follow the retreating ice edge (SEARCH, 2001; Hunt and Stabeno, 2002).

Ultimately, changes in physical forcing and the impacts on biochemical cycling influence the air-sea flux of carbon dioxide as well as the sequestration of carbon to depths at both polar regions. The Southern Ocean plays a central role in the global carbon cycle and biological productivity, and it responds to climate forcing (Sarmiento et al., 1998). Recent ecosystem studies in the Ross Sea, Antarctica, suggest that changes in ice extent and nutrient availability influence phytoplankton species' growth and the subsequent recycling of carbon in surface waters or deposition and sequestration of carbon to depth (DiTullio et al., 2000; DiTullio and Dunbar, 2003). A warming ocean may be causing the melting of the Larsen Ice Shelf in Antarctica (Kaiser, 2003; Shepherd et al., 2003), and the subsequent freshening of the surface waters is expected to alter carbon cycling in this region.

The high rate of warming on the Antarctic Peninsula—the fastest-warming region on Earth—has led to increased precipitation. When precipitation falls as snow, it can reduce the amount of bare ground that is available for penguin rookeries. Penguin populations in some areas appear to be in decline for this reason. The ectothermic ("cold-blooded") marine species of Antarctic waters, which have evolved under stable "ice bath" conditions for millions of years, may have only limited abilities to acclimate to warmer temperatures. These organisms may be especially threatened by continuing global warming.

Forcings

To understand the cause of recent changes and to project future change with confidence requires understanding of how anthropogenic forcings, natural forcings, and internal variability each contribute to recent trends. The most important anthropogenic forcings relevant to the climate of the twentieth century were an increase in greenhouse gases, an increase in sulfate aerosols, and a decrease in stratospheric ozone. The natural forcing factors included variations in solar irradiance and volcanic aerosols. Internal dynamics of the climate system alone caused the climate to be unsteady. The polar regions are prone to greater internal variability than anywhere else on Earth, owing to the presence of ice and snow and the formation of dense waters in the polar oceans. Because of the complexity of the climate system, uncertainty remains in the relative contribution of each of these to the twentieth century warming (IPCC, 2001) and to the polar environment. Nonetheless, the IPCC (2001) has stated that most of the global warming observed over the past 50 years is attributable to human activities, and there is new and strong evidence that in the Arctic much of the observed warming over the same period was also due to human activities.

Many characteristics of the recent change are consistent with those predicted by global climate models (Gregory et al., 2002). These models predict continued warming this century, especially in the Arctic, due to projected changes in anthropogenic forcing (Figure 3.4). In fact, the polar regions may already be harbingers of impending global change, but we cannot be certain because of limited understanding of the environmen-

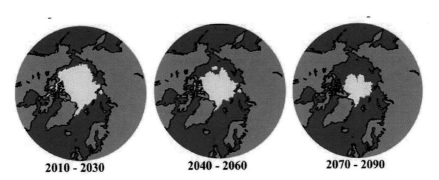

| 2010 - 2030 | 2040 - 2060 | 2070 - 2090 |

FIGURE 3-4 Projections of composite mean September Arctic sea ice changes extent based on the IPCC 2001 B2 scenario in five general circulation models: Canadian Climate Centre for Modelling and Analysis, Max-Plank Institute, Geophysical Fluid Dynamics Laboratory, Hadley Centre, and National Center for Atmospheric Research for the Arctic Climate Impact Assessment. SOURCE: ACIA, 2004.

The Sun–Earth Environment

The Sun is the source of energy for life on Earth and the strongest modulator of our physical environment. The Sun's influence spreads throughout the solar system, through photons, which provide heat, light, and ionization, and through the continuous outflow of a magnetized, supersonic ionized gas known as the solar wind. Over very long timescales (millennia and more), irregularities in Earth's orbit affect climate. One great unknown that may affect the environment on century to decadal scales (or less) is variability of the solar output. Little is known about solar variability and the potential role it may play in Earth's climate. Understanding variations, both long-term and short-term in the Sun's magnetic activity and radiative output and their couplings to Earth's space and upper-atmosphere environment, is one of the necessary conditions for distinguishing the human influence on global climate from the background of natural variability.

Solar activity, or "space weather," produces disturbances in the Earth's space environment that can adversely affect certain important technologies and threaten the health and safety of astronauts. Polar region technologies that can be affected include high-frequency communications, global positioning determinations, power systems, and pipelines. Knowledge obtained through solar and space physics research is essential to the development of means and strategies for mitigating the harmful effects of such disturbances.

Geomagnetic field lines thread through the space around Earth and physically link the northern and southern polar regions (see Figure 3.5). These polar region field lines pass close to, and indeed often form, the boundary of Earth's space environment—the magnetosphere—with the expanding solar atmosphere (the solar wind in the interplanetary medium). Variations and changes in the solar wind are thus experienced first in the polar regions and are often most visibly apparent through production of the northern and southern lights (the aurora borealis and australis) in the upper atmospheres (90 to several hundred kilometers above the Earth's surface) of the polar regions. Disturbances by the expanding solar atmosphere can often be felt throughout Earth's space environment.

Major challenges in understanding the Sun and the heliosphere include (1) understanding the structure and dynamics of the Sun's interior, the generation of solar magnetic fields, the origin of the solar cycle, the causes of solar activity, and the structure and dynamics of the solar corona, and (2) understanding the interactions of the expanding solar atmosphere with Earth's magnetosphere.

tal system and the inaccuracy of models. A systemwide assessment of the polar regions and an expansion of monitoring networks are necessary to improve our understanding of the character, mechanisms, and impacts of change and to improve our ability to predict future change.

LESSONS FROM PAST CHANGE

Past change can provide a context for evaluating modern observations of change. For example, proxies of surface temperature, such as from tree rings, pollen, sediments, and ice cores, indicate that the surface warming of the twentieth century averaged over the Northern Hemisphere (of about 0.6°C over land and ocean) is likely to have been the largest of any century during the past 1,000 years (IPCC, 2001). A similar analysis

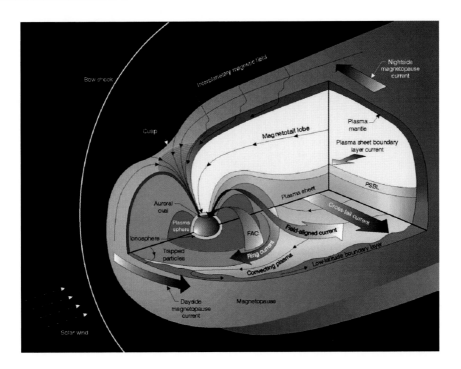

FIGURE 3-5 Artist's conception of Earth's magnetosphere, the volume of space around Earth dominated by the geomagnetic field and populated with plasmas of both ionospheric and solar wind origin. The ionized gases that populate the magnetosphere are remarkably dilute: the densest magnetospheric plasma is 10 million times less dense than the best laboratory vacuum. Nevertheless, the motions of these highly tenuous plasmas drive powerful electrical currents, and during disturbed periods, the Earth's magnetosphere can dissipate well in excess of 100 billion watts of power—a power output comparable to that of all the electric power plants operating in the United States.

of proxy data from high northern latitudes in combination with climate modeling has shown that the warming during the twentieth century was most likely a result of increased levels of greenhouse gases, as opposed to solar variability and volcanic aerosols (Overpeck et al., 1997).

Large-Scale Environmental Change

The environment has undergone rapid and dramatic shifts in the past, and there is no reason it could not experience similar changes again in the future (NRC, 2002). Figure 3.6 shows temperature and snow accumulation since the tail end of the last ice age from a Greenland ice core, including the large change in temperature in the transition from the last ice age toward modern values. Most of the change since the last

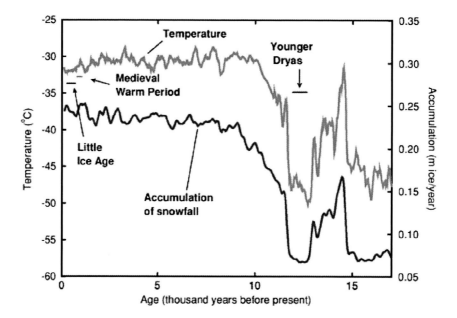

FIGURE 3-6 Climate changes in central Greenland over the last 17,000 years. Reconstructions of temperature and snow accumulation rate show a large and rapid shift out of the ice age about 15,000 years ago, an irregular cooling into the Younger Dryas event, and the abrupt shift toward modern values (Cuffey and Clow, 1997; Grootes and Stuiver, 1997). The 100-year averages shown somewhat obscure the rapidity of the shifts. Most of the warming from the Younger Dryas required about 10 years, with 3 years for the accumulation-rate increase. A short-lived cooling of about 6°C occurred about 8,200 years ago. Climate changes synchronous with those in Greenland affected much of the world. SOURCE: NRC, 2002.

peak of the ice age, about 12,000 years ago, in a period known as the Younger Dryas, occurred within a span of a few decades. Yet much smaller changes of only a degree or two, which are known to have occurred in between the so-called Medieval Warm Period and the Little Ice Age, were sufficient to cause Scandinavian settlers to colonize Greenland during the warm periods and then to abandon those farms during the cold. Many past examples of societal collapse involved rapid change to some degree (NRC, 2002).

Natural records preserved in sediments, caves, and ice cores show that the climate system has crossed thresholds that resulted in abrupt changes, where a small change in one part of the system resulted in a sudden and large response in another (NRC, 2002). For instance, ice core records indicate that Greenland warmed by about 10°C in as little as a decade more than a dozen times during the last glacial cycle (Severinghaus et al., 1998; Alley et al., 1993). These abrupt warmings were followed by a slow cooling over roughly 500 years (Figure 3.6), with about 1,500 years between abrupt changes (Rahmstorf, 2003). These timescales suggest that the ocean and/or continental ice sheets might be key players. Although the precise nature of the threshold mechanism is unknown, these players are active in our climate system today. If Earth continues to warm beyond natural rates, abrupt change could be in our future.

Paleobiology

Just as the chemistry of polar ice reveals much about the environments of the past, the layered chemical and biological histories preserved in the ice also preserve evidence of past ecosystems. Ice coring studies have shown that in some cases metabolically active microbes may exist in small liquid water veins in solid glacial ice (Price, 2000), or organisms may be cryopreserved and metabolically inactive within the ice; viable microbes thousands of years old have been found in ice cores (Doran et al., 2003). Genomic methods provide a tool for characterizing the identity of organisms and organism remains and may enable new links between climate and biological adaptation, as well as insights into conditions for life on other planets.

From October 1997 until October 1998, scientists based on a ship frozen into the Arctic ice near the North Pole conducted experiments to better understand changes in the ice, ocean, and atmosphere. The project, known as SHEBA (Surface Heat Budget of the Arctic), was supported by the National Science Foundation. SOURCE: SHEBA Project Office.

As part of the SHEBA ice station project, Jackie Richter-Menge and Bonnie Light read ice thickness gauges as the ice melted in August 1998. SOURCE: SHEBA Project Office.

Forcings

Paleo evidence from the last few ice ages provides a rich picture of the climate variability on multidecadal to millennial timescales, which is fundamentally driven by internal dynamics of the environment subject to variability in the total solar irradiance and volcanic aerosols. It shows how the climate is capable of behaving in the absence of anthropogenic forcing. We must look on longer timescales to understand how climate can respond to more substantial natural forcings. Paleo data indicate that feedbacks in the polar regions have been responsible for enormous environmental change owing to alterations in the Earth's atmospheric composition and incoming solar distribution, the buildup of ice sheets, and continental drift (Crowley and North, 1996). Figure 3.7 shows

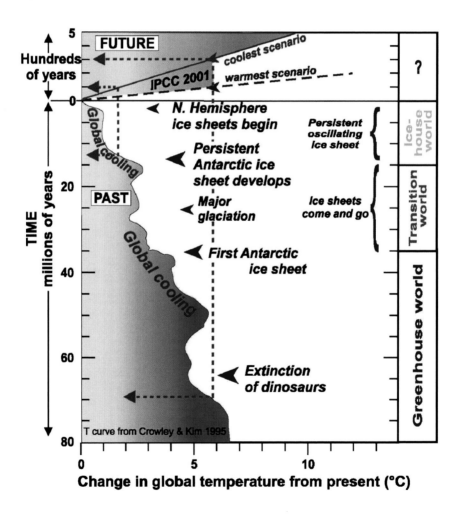

FIGURE 3-7 Variation in Earth's temperature during the past 80 million years, based on reconstructions from deep-marine oxygen isotope records. SOURCE: Barrett (2003).

the history of global temperatures for the past 80 million years. Major puzzles remain regarding why glaciations were intermittent and what caused major warmings and coolings on top of the very long global cooling of the past.

While exciting discoveries have prompted a great deal of recent theoretical advancement, the fundamental mechanisms driving variability on a variety of timescales are still not well understood. Expanding data-gathering efforts to provide new proxy records on past climates would help improve our understanding of recent discoveries. Added spatial coverage would allow us to identify modes of climatic variability, which may shed light on the basic physics driving these modes and their relevance to the modern environment and societies.

Unraveling the Secrets within Polar Ice

When Richard Alley thinks about working at the poles, he thinks about sitting in the midst of endless snow, in the middle of Greenland, 200 miles from anywhere, and having an Arctic fox trot into camp. He thinks of white snow and blue snow, white clouds and blue skies. He thinks of black rocks along the coast and red coats on people; of musk, seals, and penguins, and snow crunching underfoot. He thinks of the aurora.

"The aurora is wonderful. It is the bioluminescence of a blooming sea spread above rather than below."

As a professor of Geosciences at The Pennsylvania State University, Alley has been on six expeditions to Greenland and three to Antarctica, and he's worked in Alaska as well. Primarily, he investigates the role of glaciers and ice sheets in the climate system, including the flow of ice and possible effect on sea-level change, and records of climate change preserved in ice cores. He does this, he says, "because, it's interesting, it's fun, and it will help people."

In the 1990s, Alley and his colleagues participated in drilling two miles into Earth's ice in Greenland. There they found atmospheric chemicals and dust that enabled them to glimpse such phenomena as wind patterns and precipitation for the past 110,000 years. The cores revealed a sweeping story of climatic history as clear as that found by reading any book.

"The cores tell us of the great ice-age roller coaster that cooled the Earth over tens of thousands of years and grew ice across almost one-third of the land—and then melted most of that ice away," he explains. "More surprisingly, although the warming from the ice age took 10,000 years, about half of the warming in Greenland happened within about 10 years. At the same time, very large and rapid climate changes affected much of the Earth. Similar abrupt climate changes happened repeatedly through the ice age and even into the warmth of the last 10,000 years— almost like a climatic bungee-jumper on the ice-age roller coaster."

Richard Alley on the Matanuska Glacier in Alaska. SOURCE: Todd Johnston.

Currently, Alley is part of a team finishing up a study of an ice core from West Antarctica. They are planning for a second, deeper core there. The Antarctic projects are intended to find the southern hemisphere equivalent of the Greenland coring. By comparing climatic history in the northern and southern hemispheres, scientists will be able to better understand what happened in the past, which should help in predicting global trends in the years ahead.

continued

Alley's findings from the Greenland ice cores were published in *The Two-Mile Time Machine: Ice Cores, Abrupt Climate Change, and Our Future,* in 2000. The book shows how wiggles in Earth's orbit, changes in ocean currents, volcanic eruptions, variations in greenhouse gases from humans and nature, and more, have combined to provide the fascinating climatic history of Earth. It also suggests how humans are becoming more and more important in controlling the future of the climate, and how abrupt climate changes might affect our choices about what to do in the future.

Humans are changing the atmosphere in many ways, according to Alley. "Our leaded gasoline dirtied the snow of Greenland until we decided to get the lead out. Cleaner snow now shows our success. But air bubbles in the ice reveal that CO_2 and other greenhouse gases, mostly from our activities, are building up to levels not seen for a very long time. Historians and physicists agree that this will warm the planet."

According to Alley, mountain glaciers already are melting as the world warms, and their water is swelling the oceans. Too much warming could melt the great ice sheets in Greenland and Antarctica, with much more impact on the coasts. If Greenland were to melt, sea level would rise a bit more than 20 feet, which without sea walls, would put Miami, Florida, mostly under water. If all the ice sheets were to melt, sea level would rise 200 feet or slightly more, which would put Florida under water and the coast up in Georgia.

Melting ice puts fresh water into the nearby oceans, which might change the way the oceans circulate. If this occurs, it will not cause a new ice age, an end to civilization, or a really dramatic disaster movie—but it could cause cooling in some regions, warming in others, and shifts in rainfall and droughts, with large enough impacts to matter to many people.

"We know that days are usually warmer than nights, summers usually warmer than winters, but we still check the weather forecast, because more information really is useful," says Alley. "Similarly, decades with high carbon dioxide usually are warmer than those with much lower carbon dioxide, but we have a lot of work to do to make climate forecasts as useful to humanity as weather forecasts are now. Our studies of ice are part of that important work. I can't wait to get back to work!"

4

Exploring Scientific Frontiers

"Never lose a holy curiosity."

ALBERT EINSTEIN

Exploration of the unknown has been a vital part of humanity's interaction with the polar environment since the first people lived and hunted in the Arctic thousands of years ago. Heroic tales of exploration are filled with adventurers from the polar regions—Amundsen, Byrd, Hensen, Nansen, Scott, and Shackleton are but a few of the many famous explorers. In earlier International Polar Year (IPY) and International Geophysical Year (IGY) research programs, science-driven exploration of new geographical regions was a major activity. In the IPY 2007-2008, only limited regions of Earth's surface, such as parts of East Antarctica, remain to be explored in the traditional geographic sense. But vast regions of the seafloor and ocean remain unexplored, and the least explored of the oceans are the Arctic and Southern. New frontiers and challenges loom as exploration activity takes advantage of new disciplines and technologies that were unknown in the previous IPYs and the IGY. Modern exploration occurs on a tremendous range of spatial scales, from the previously inaccessible realms of the genome to Earth's mantle, core, and magnetosphere. Fascinating discoveries can be expected.

EXPLORING LIFE IN THE POLAR REGIONS

The extremes of temperature, the ice cover, and the marked seasonal variation in day length that characterize the polar regions confront organisms with a challenging set of environmental conditions. Yet the apparent "hostility" of the Arctic and Antarctic environments has not precluded the evolution of complex ecosystems whose constituent species have adapted in novel ways to the extreme physical conditions they face. Biologists have much to learn from the study of all types of polar organisms, including microbes, the most abundant yet least understood group of species; insects and plants, which have remarkable

41

abilities to either avoid freezing or to survive the formation of ice in their body fluids; and hibernating mammals, whose capacities to survive near-freezing body temperatures may provide new insights into biomedically important issues such as cold-preservation (cryopreservation) of tissues and whole animals. Many new species of polar organisms remain to be discovered, as shown by the recent finding of new species of invertebrate and vertebrate marine life through the internationally organized Census of Marine Life project.[1]

The biology of the polar regions thus represents a fascinating frontier for exploration, where rewards in both basic and applied science are waiting (NRC, 2003a). What follows is an overview of some types of exploratory activities in polar biology that seem especially likely to provide important new knowledge about polar ecosystems and the adaptations that make life possible under the extreme physical conditions of the polar regions. Many of the key questions in polar biology can only now be answered because of the development of new technologies, ranging from those of genomics to cold-hardened, remotely operated vehicles (ROVs). The tools are now at hand to explore the biota of the polar regions as never before possible.

Microbiology

We currently know relatively little about the numerically dominant species in the polar regions—the microbes (NRC, 2003a). These diverse species, which comprise members of the Eukarya, Bacteria, and Archaea domains, are involved in virtually all biogeochemical transformations in terrestrial, freshwater, and marine ecosystems of the polar regions. Thus, until we thoroughly explore the microbial world at the polar regions, we will lack the basis for a comprehensive understanding of the functions of polar ecosystems and their susceptibility to climate change and pollution. Our current state of knowledge—or ignorance, to be more accurate—about polar microbes is illustrated by the fact that we are largely unable to answer the following two basic questions: "What types of microbes are present in polar ecosystems?" and "What types of metabolic activities and ecological transformations do polar microbes carry out?"

Our inability to provide detailed answers to these fundamental questions largely reflects past technological limitations to the study of microbial life, limitations that newly developed genomic technologies have mostly eliminated. As discussed in NRC (2003a), genomic methodologies, which are defined as "the study of the structure, content, and evolution of genomes, including the analysis of expression and function of both genes and proteins," allow identification of species and elucidation of the types of functions their genes and proteins enable them to perform. For example, the types of genomic techniques used by law enforcement agencies in "molecular forensics" can be applied to identifying the diverse types of microbes present in aquatic and terrestrial environments. Other genomic methodologies, such as those that employ bacterial artificial chromosomes, allow characterization of the genes that underlie metabolic

---

[1]The Census of Marine Life (CoML) is an international effort directed toward assessing and explaining the changing diversity, distribution, and abundance of marine species on a global scale. CoML programs that specifically target the Arctic and Antarctic are currently in development. At the U.S. national level, the National Oceanic and Atmospheric Administration's Ocean Exploration program has already mounted such expeditions to the Arctic's deep basins to search for novel organisms in the sea ice, in the water column, and on the seafloor. Both initiatives hope for significant involvement in IPY activities.

"Black smokers" are fissures in the sea floor that emit scalding hot vent waters, whose temperatures may reach 750°F (400°C). In contrast, the normal ocean bottom water around the smokers is close to 35°F (2°C). While no organisms live in the hottest water coming from the smokers, small worms and crabs snuggle into the warm walls of the chimney. Recent investigations revealed a surprising amount of "black smokers" and associated hydrothermal systems along the Arctic spreading ridges. Biologists are puzzled as to how vent animals adapt to such high pressure and warm temperatures. Microbiologists are especially interested in cultivating "extremophiles," microbes that can withstand extremely high temperatures and high levels of heavy metals. SOURCE: Jim Childress, University of California, Santa Barbara.

function. Thus, these new DNA-based methods now provide microbiologists with tools to determine what microbes are out there and what roles they play in ecosystems. We are poised to explore the world of polar microbiology in wholly new ways, and this work will open up important new understanding of the pivotal role that microbes play in all polar ecosystems.

Biotechnology and Biomedicine

The remarkable abilities of plants and animals native to polar regions to withstand extremes of low temperature and, in many cases, wide ranges of temperature represent a promising biological frontier for exploration. Genomic technologies offer the prospects of discovering important new mechanisms of adaptation that polar organisms have "invented" during their evolution under physical extremes and of developing ways to exploit these adaptations in biomedical and biotechnological research (NRC, 2003a).

Of particular interest in the context of biomedical research are the abilities of many polar organisms to either avoid freezing or to withstand the formation of ice in

Almost a thousand Adelie penguins live in the Cape Royds rookery on Ross Island in the Antarctic. Rookeries serve as nursery grounds for the young birds. In the background is Mount Erebus, a smoking active volcano. SOURCE: George Somero, Stanford University.

their body fluids. Small mammalian hibernators such as ground squirrels allow their core body temperature to fall to approximately −2°C. Yet these animals do not freeze and they rapidly return to a metabolically active state when the body's heat-generating mechanisms are activated. Our ability to develop improved methods for low-temperature storage of biological materials—ranging from isolated cells to intact organisms—may depend on acquiring a detailed understanding of the unique physiologies of hibernators. Many polar insects and plants attain even lower cell temperatures in their terrestrial habitats, yet their cells remain ice-free because of the accumulation of antifreeze compounds. Understanding the fundamental mechanisms of freezing resistance will have broad technological applications, ranging from agricultural science (e.g., the design of freeze-resistant crops) to biomedicine (e.g., development of improved cryopreservation techniques). Some polar animals and plants do experience the formation of ice in the extracellular fluids, yet these species appear to be undamaged by this type of freezing. The factors that regulate freezing of extracellular water and protect against damage from ice formation may allow further advances in cryotechnology and biomedicine. The new genomic technologies that have been developed largely in the context of biomedical research offer a remarkable new set of tools for polar biology, tools that will allow contributions to be made in both basic and applied research.

Polar Ecosystems

Who could have predicted that the study of polar ecosystems would reveal fishes that, unique among vertebrates, lack red blood cells; hibernating mammals whose body temperatures plummet below 32°F (0°C) in winter; algae, living within ice and quartz-containing rocks, that may be metabolically active for only hours each year; fish whose blood remains in the liquid state at subzero temperatures because of the presence of novel biological antifreeze proteins; insects that are able to crawl on the snow at temperatures of –58°F (–50°C), and large subglacial lakes, isolated from the rest of the biosphere for many millions of years, that may hold a variety of ancient forms of life?

Polar Ecology

Inasmuch as new technologies offer the means to explore the genomes and physiologies of polar species, new tools are becoming available that allow novel types of exploration of large and complex terrestrial and aquatic ecosystems in the polar regions (NRC, 2003a). Newly developed electronic tagging technology enables the movements

A wide variety of animals thrive in the ice-cold waters of the Antarctic Ocean and play key roles in the marine food web. Here, the notothenioid fish *Trematomus bernacchii* swims past a sea spider, *pycnogonid*, and a sea star. SOURCE: George Somero, Stanford University.

Surviving Winter at the Polar Regions

Evolutionary processes have created many biological communities that are as stunning in their beauty and complexity as they are unexpected by and novel to biological scientists. The Arctic ground squirrel and the black bear are two of many such models of polar species with mysterious mechanisms for surviving extreme environments. In this case the animals enter a state of suspended animation called hibernation. Understanding how this happens may one day provide huge medical benefits that help alleviate human suffering.

The black bear, for example, survives polar winters without eating or drinking for as long as six to eight months by lowering its metabolism. Unlike humans, however, who experience significant bone loss if immobilized for extended periods of time, the bear maintains bone mass. By unraveling the biology behind how bone is preserved, biologists will be one step closer to preventing osteoporosis in chronically bed-ridden patients and others prone to this condition.

In contrast to the black bear, whose body temperature drops by about 9°F (5°C) during hibernation, the Arctic ground squirrel's body temperature plummets as much as 72°F (40°C) to attain a below-freezing core body temperature of about 28°F (−2°C). Interestingly, blood flow to tissues during this period is reduced by as much as 98 percent for up to three weeks, and yet no tissue damage occurs. If biologists can find the biochemical pathways that afford cells protection from low blood flow, they may be able to protect individuals from injury due to strokes and heart attacks when reduced oxygen is available to cells (Becker et al., 2002; Boyer and Barnes, 1999).

of animals to be tracked remotely over long distances, providing important information about the sites at which species feed and reproduce and the physical conditions of the environment (temperature and light levels). Satellites may provide an important link in these tracking studies, further emphasizing the role of these space-based platforms in polar research. New, cold-hardened models of remotely operated vehicles being deployed in polar oceans are providing rare insight into the unknown biology of these seas, finding new species of life and allowing studies of modes of adaptation by these organisms to these extreme environments. Genomic techniques can be used to analyze the genetic structures of populations of plants and animals, revealing whether genetically distinct populations occur in habitats with different environmental characteristics. And, as discussed above, the recently acquired abilities to determine the types of microbes present in an ecosystem and to elucidate the physiological activities of these important organisms open up an especially important new avenue of exploration in polar ecosystems.

EXPLORING NEW REGIONS

Subglacial Lake Environments

Recent discoveries show that buried under miles of Antarctic ice are subglacial lakes ranging in size from Lake Vostok (Figure 4.1), a body the size of Lake Ontario, to shallow frozen features the size of Manhattan. Over 100 lakes, both shallow and deep, have now been identified, suggesting that the subglacial environment is a previously unrecognized immense, possibly interconnected, hydrological system. The extent and degree of interconnection among the lakes are unknown. These recently discovered

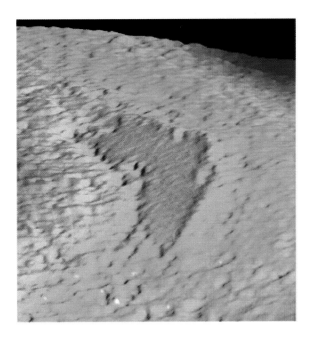

FIGURE 4-1 Perspective view of the ice surface above subglacial Lake Vostok based on ERS-1 radar altimetry data. The ice above Lake Vostok shows as the flat, featureless region in the center of the image. SOURCE: Michael Studinger, Lamont-Doherty Earth Observatory of Columbia University.

subglacial environments formed in response to the complex interplay of tectonics and topography with climate and ice sheet flow over millions of years.

The temperatures and pressures of subglacial lakes are relatively moderate, similar to the environment of the deep oceans. However, subglacial environments are unique hemispheric-scale laboratories found nowhere else on Earth. Sealed from free exchange with the atmosphere for 10 million to 35 million years, subglacial environments are the closest Earth-bound analogs to the icy domains of Mars and Europa (Siegert et al., 2001).

Tantalizing evidence from studies of the overlying ice sheet indicates that unique life-supporting ecosystems may be locked within these subglacial environments (Priscu, 2002; Priscu et al., 2002). Such life forms must be adapted to the temperatures and pressures akin to the deep ocean, as well as the extremely slow delivery of nutrients from the overriding ice sheet. These subglacial environments provide an unparalleled opportunity to advance our understanding of how climatic and geological factors have combined to produce a unique and isolated biome that may be occupied by yet unknown microbial communities.

Subglacial lake exploration poses one of the most challenging scientific, environmental, and technological issues facing polar science today. Many engineering issues need resolution, sensors and instrumentation must be developed, and environmental assurance must be demonstrated before scientific exploration is feasible. IPY 2007-

2008 will spawn new technologies and partnerships that can accelerate solving these challenges, leading to fundamental discoveries about life and climate on Earth and the associated implications for other planets.

Subglacial Lands

The lands beneath the ice sheets of Antarctica and Greenland remain poorly understood and largely unexplored, yet the geological and hydrological characteristics of these subglacial regions are vital for understanding ice sheet development. The nature of the underlying bedrock is a crucial boundary condition that defines the stability of the ice sheet to climatic changes. The subglacial topography is key to ice sheet models in part because the distribution of highlands strongly defines when, where, and how glaciation initiates (DeConto and Pollard, 2003). Much of the subglacial lands in Antarctica and Greenland remain unexplored, so our understanding of how these regions became icebound remains unconstrained.

Major regions of Antarctica that are crucial to deciphering the intertwined geodynamic/climatic history puzzle remain to be explored for the first time. For example, the Gamburtsev Mountains in East Antarctica cover an immense region the size of Texas, yet detailed topography and peak elevation of the mountains remain matters of conjecture. Climate models suggest that the high elevation of these mountains was crucial in localizing the first Cenozoic ice sheets that formed 34 million years ago (Figure 4.2). This onset of glaciation affected the entire Earth, as global climate changed from the hothouse world of the early Cenozoic to the more recent world in

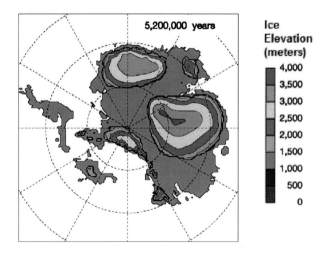

FIGURE 4-2 Glacial ice elevation in Antarctica 5.2 million years ago following a decrease in the level of carbon dioxide in the atmosphere, from the coupled climate-ice sheet model of DeConto and Pollard (2003). This model simulates the glaciation following global cooling 34 million years ago. Glaciers first formed in the East Antarctic highlands and moved outward to encompass the entire continent. SOURCE: DeConto and Pollard (2003).

which whole continents are covered in ice. Learning the geological history of the Gamburtsev Mountains enables understanding of how their elevation may have affected climate evolution through the Cenozoic. Clarification of the processes involved in Gamburtsev Mountains uplift will place constraints on the origin of paleotopography of East Antarctica and its possible role in localizing initial glaciation.

Important questions remain concerning the relationship between West Antarctic tectonics and the stability of the West Antarctic Ice Sheet. Previous radar imaging work suggests possible ongoing volcanism and active continental extension in the West Antarctic Rift System (Bell et al., 1998; Anandakrishnan et al., 1998; Blankenship et al., 1993). However, little is known about this volcanism, and the possible role of elevated heat flow from the continental extension in enhancing flow in the overlying ice streams has not been evaluated. Other research topics in need of investigation involve subglacial geology and heat flow in Greenland, via the northeastern Greenland Ice Stream and the processes involved in sub-ice-shelf environments, melting, and circulation. Exploration of subglacial lands and environments will provide much-needed constraints on the evolution of climate-ice sheet systems.

Ocean Basins

The polar ocean basins beneath sea ice represent vast unexplored regions that can now be studied with modern technology, such as remotely operated vehicles and autonomous underwater vehicles. These advances will allow biologists to investigate questions about variations in ecosystems between the ice edge and regions deep within the ice pack and the types of seasonality found beneath floating ice shelves. Recent ocean exploration studies in the Arctic have found new species in both the column and the sediments, emphasizing the "spacelike" unknown environments still common in the extreme environments of Earth. How organisms have evolved to adapt to cold isolated environments will provide important understanding of the processes necessary for life in this extreme environment. In addition, subsea ice exploration can clarify the seasonal changes in both continental shelf dynamics and shelf break upwelling in polar marine seas. These have been argued to result in a major impact on uptake of carbon dioxide as well as global thermohaline circulation.

The Gakkel Ridge in the Arctic Basin is the slowest spreading midocean ridge on Earth, yet study has been limited to only a few submarine and icebreaker expeditions from 1995 to 2003. Study of the Arctic Basin ridges will provide important end-member constraints on ridge processes that vary with spreading rate. The few previous expeditions suggest that the Gakkel Ridge has the thinnest-known oceanic crust. However, some sites display abundant hydrothermal and volcanic activity, and the ridge exhibits long-lived segmentation controlled by mantle processes. These long-lived hydrothermal ecosystems may have been cut off from the rest of the oceanic ecosystem for a long time, since the ridge segments are isolated. Thus, the deep-water vent fauna ecosystems may contain a large number of endemic species and provide constraints on the genetics and evolution of seafloor organisms.

The tectonic history of the Arctic Basin is relevant for understanding past ocean circulation and climate, yet very little is known about it. Exploration of this ocean floor will clarify the geological history of these regions and allow reconstruction of Arctic tectonics and its influence on ocean circulation. The tectonic history of ocean gateways, which allow passage of warm or cold currents between oceans, is useful for understanding climate in both Arctic and Antarctic regions. For example, Fram Strait, between Svalbard and Greenland, is the only deep-water gateway between the Arctic basin and the global oceans and the date of its formation is unknown. Similarly, constraints on the opening history of Australia-Antarctica and South America-Antarctica gateways will allow better understanding of the onset of the Antarctic Circumpolar Current and its effects on climate and biological evolution.

Earth's Deep Interior

Earth's deep interior holds many clues to understanding the planet's processes and history. The recent development of imaging techniques such as seismic tomography now allows important details of Earth's interior beneath the polar regions to be resolved for the first time. For example, the surface elevation of East Antarctica is a key factor for localized glaciation. The reason for this anomalous elevation is unknown and may be the result of compositional, thermal, or dynamic buoyancy in the mantle. Detailed seismic imaging of the mantle will resolve this question.

Exploring Earth's interior will also help constrain coupled climate-icesheet models that require knowledge of basal heat flow and mantle viscosity. Heat flow is an important parameter since ice viscosity is strongly temperature dependent and mantle viscosity controls the rate of terrestrial response to ice loading. The temperature of the upper mantle, estimated from seismic velocity maps (Figure 4.3), suggests a difference

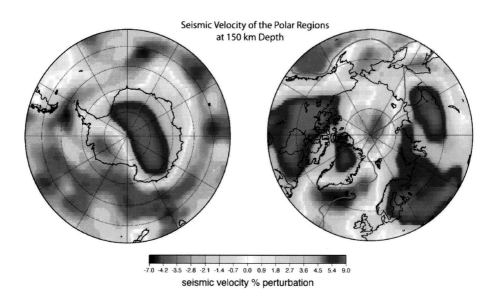

FIGURE 4-3 Map showing seismic velocities at a depth of 150 kilometers beneath the polar regions. Red colors denote slow velocities associated with hot and upwelling regions in the mantle, and blue colors denote fast velocities resulting from cooler temperatures and down-welling regions. The warmer mantle regions are associated with magma production, polar volcanism, rapid postglacial rebound, and increased heat flow beneath glaciers. SOURCES: Ritzwoller et al. (2001) and Shapiro and Ritzwoller (2002).

of about 600°C between East and West Antarctica at a depth of 150 kilometers. These temperature differences have a profound effect on heat flow and mantle viscosity influencing ice sheet models.

Imaging the Earth's interior can also provide constraints on the origin and spatial distribution of magma sources for polar volcanism because melt-producing regions are characterized by low seismic velocity. Mapping the distribution of melt-producing regions and mantle flow patterns beneath the Iceland hotspot and the slow oceanic spreading centers of the Arctic basin is possible. The increased magmatism and mantle flow associated with the Iceland hotspot may have had a large effect on the tectonic and climatic evolution of Greenland and the Arctic basin. Investigation of mantle anomalies beneath the Gakkel Ridge may provide the first direct evidence of the distribution of melt production regions along an ultra-slow-spreading center.

Exploring the core of our planet is essential for understanding the initial differentiation of the Earth, the Earth's thermal history, and the physics and variability of the Earth's magnetic field. The inner core, which solidified from the liquid outer core as the Earth cooled, shows seismic anisotropy approximately aligned with the Earth's rotation pole. Thus, seismic phases traveling along polar paths show much smaller travel times than the same phases traversing the inner core along equatorial paths. This anisotropy results from the preferential orientation of iron crystals, which may be caused by alignment due to convection patterns in the core. Observations of polar path seismic waves also suggest that the inner core rotates slowly relative to the Earth's

The submarine USS Hawkbill broke through the Arctic ice in April 1999 while on a science mission conducted with the National Science Foundation. During the cruise and associated ice camp, researchers made a variety of physical, chemical, and biological measurements and military personnel gained important experience in ice operations. SOURCE: US Navy.

mantle and surface. More detailed observations from seismographs in polar regions are essential to constrain the distribution of crystal alignment in the inner core and for understanding the dynamics of the core and changes in the Earth's magnetic field.

Sun and Solar-Terrestrial Connections

At first glance, studies of the Sun and Earth's solar-terrestrial environment would not appear to be "new frontiers" as envisioned in the International Polar Year. The IGY provided the major discovery of the Van Allen radiation belts and a revolution in understanding—continuing to the present—of the Earth's space environment. The Sun, being the closest star and responsible for all life on Earth, continues to be studied around the globe. But in reality, much remains to be learned about the Sun and the interactions of its highly variable photon, plasma, and particle emissions, which are the key "upper" boundary conditions to all processes at work in the polar regions. A better understanding of the Sun and solar variability is necessary to comprehend how "natural" variations affect polar phenomena and human existence. Therefore, it is essential that IPY 2007-2008 be deeply involved with ongoing and creative new studies of the Sun and its manifold influences on the Earth.

Magnetic field lines stretching out from the polar regions interact with the flowing and variable solar wind, transferring electromagnetic and charged particle energy to the upper atmosphere of the polar caps. The portions of such energies that may be responsible for such important polar phenomena (e.g. noctilucent clouds) are completely unknown today (NRC, 2003b). Variabilities in the emission of solar photons over all wavelengths—the so-called solar constant—affect the polar regions and global climate in ways that are only beginning to be studied through models and simulations. Global cloud cover data, including in the polar regions, which are important for models and which can be affected by solar emissions and their variability, are almost absent from databases of the polar environment. Except for the past 10 years or so, actual data on solar variability that need to be incorporated into models are largely by proxy from studies of polar and glacial ice sheets, ocean sediments, and other terrestrial sources. New spacecraft measurements made by improved instrumentation are beginning to monitor, with good precision, the solar photon output.

But the underlying physics of the driving source(s) on the Sun—largely via magnetic fields—for its activity and variability is poorly understood. This lack of understanding, and the acquisition of the essential new data, will not magically be remedied or accomplished during the next IPY. What the IPY will be able to accomplish in terms of the Sun-Earth connection is to ensure that these "above the tropopause" inputs to the polar regions will be intensively examined and incorporated into studies of natural polar change and that the means for continuing detailed studies of the Sun and its interactions with Earth, so necessary for polar science, will be a legacy of IPY 2007-2008.

Studying Ecology From Pole to Pole

The shrill whining of his chainsaw shattered the silence of the Antarctic lake as Charles Goldman worked his way toward the algae beneath the frozen coastal ice. Distinguished professor of limnology, or aquatic ecology, at the University of California, Davis, Goldman studies inland waters such as lakes, rivers, and estuaries—literally from pole to pole.

His concentration is on primary productivity and limiting factors, or nutrients, in inland bodies of water. In other words, he investigates whether plants and animals have everything they need for good health and reproduction—or whether there is a single factor that may be limiting the growth, abundance, or distribution of the population of a particular plant or animal species.

In this case, Goldman found

Lake Bonney in Taylor Dry Valley, Antarctica, across McMurdo Sound from Ross Island. Dry valley lakes like Lake Bonney are permanently ice covered. Underneath several feet of ice, however, rotifers and other tiny animals swim in an aquatic niche of liquid water. SOURCE: George Somero, Stanford University.

the algal plants he was looking for, "the algae were concentrated under the ice at temperatures below freezing." Incredibly, they were still carrying out photosynthesis at a slow rate.

"The Antarctic is a wonderful place to work," Goldman says. "It is so unspoiled. This is especially important for a limnologist because so few organisms are present. You are dealing with simple ecosystems here, so it's easier to unravel relationships between organisms and their environment. It's quiet too. It's the quietest place on Earth."

Goldman's entire career has been centered on this kind of water research, and he has worked in countries around the world. In Antarctica, for example, Goldman's work has been concentrated on shallow coastal lakes that thaw in the Austral summer. He has also worked on the permanently frozen Dry Valley lakes and has a current interest in Lake Vostok, a newly discovered subglacial lake about the size of Lake Ontario that is under miles of Antarctic ice. Subglacial lakes are of keen interest because no one knows if they hold life or understands the biochemical processes that must exist to sustain life without sunlight, with low nutrient inputs, and in temperatures below freezing.

Goldman's Arctic work includes research in Swedish Lapland, where he introduced more than 6,500 Lake Tahoe crayfish to reestablish the Scandinavian crayfish industry. In Alaska he has worked with the U.S. Fish and Wildlife Service to increase the primary productivity of Sockeye salmon nursery lakes.

Goldman says that the adult fish lay eggs in the gravel. Then they die and pile up in mounds at the mouths of streams. The carcasses decompose and, as they do, they fertilize the streams. Come spring, when the baby fish emerge from their egg sacks, they have the nutrients they need to survive as they arrive at the lakes.

"Most of the mortality of salmon occurs at the freshwater stage," Goldman says. "Anything we can do to increase survival of the fish at this point can make a huge difference in the return you get back from the sea." As a result of the findings of Goldman and others, Sockeye salmon lakes have been fertilized and productivity of the salmon has increased.

"As president of the newly formed World Water and Climate Network, I am keenly interested in continuing polar exploration," says Goldman. He views the International Polar Year 2007-2008 as an opportunity to turn our attention to the global availability of water, the most valuable resource on Earth.

"The polar regions will be increasingly important in years to come," Goldman predicts. "It is both of national and international self-interest to keep our Arctic and Antarctic programs in place and pollution-free."

5

Technology to Enable Innovative Observations

"Beauty and grace are performed whether or not we will or sense them.
The least we can do is try to be there."

ANNIE DILLARD

Observations of many significant components of the polar regions remain extremely limited and nonstandardized, due to the small, scattered human populations, limited scientific infrastructure, and inherent difficulties of working in cold, remote environments year-round over sustained periods of time. Additionally, observation infrastructure and records are being lost, reduced, or eliminated in some countries, further restricting our ability to understand these complex regions. The International Polar Year (IPY) can make a major contribution to science and society by using the 24-month window during 2007-2008 to intensively observe and explore the polar regions at greater spatial and temporal resolutions, by undertaking international data rescue efforts, and by setting in place an observation network to enable ongoing observations of the polar regions in the decades to come. Innovative technologies will be instrumental for analyzing the myriad observations collected during the IPY and for disseminating IPY research to the broader public.

OBSERVING THE POLAR REGIONS DURING IPY 2007-2008

The efforts of the IPY 2007-2008 to assess the changing polar environments, explore new frontiers, implement interdisciplinary observatories, understand the human-environment dynamics, and build new connections between the public and the polar regions will require implementation of innovative technologies. In the temperate regions the past 50 years have brought the tools of science to a new level—measurement and observing technology, computing capacity, miniaturization, and a host of other advances. Measurements and techniques are possible now that were only seen in science fiction novels during the International Geophysical Year (IGY) in 1957-1958. The availability of innovative satellite data and drilling techniques, as well as access to suborbital platforms and subsea vessels, will be vital for conducting intensive field studies during the IPY 2007-2008 and beyond.

Satellites first appeared as observational platforms during the IGY, and their initial instrument payloads immediately made major contributions to the success of IGY. Satellite-borne remote sensors of the Earth's surface and space environment, as well as of the Sun, are continuing to evolve and become more sophisticated, and they will undoubtedly play an even more important role in the next IPY. Existing satellites obtain information across much of the electromagnetic spectrum and provide high spatial and temporal resolution data over the polar regions.

The rapidly approaching IPY precludes the inclusion of heretofore unplanned missions, but a number of planned U.S. missions with specific utility for the polar regions will be available in 2007-2008 (Table 5.1). The National Aeronautics and Space Administration's (NASA) "A-Train" satellite formation consists of two major Earth Observing System (EOS) missions, three Earth System Science Pathfinder missions, and a French Centre National d'Etudes Spatiales mission flying in close proximity. This carefully planned formation of polar orbiting satellites allows for synergy, meaning that more information about the Earth and its polar regions is obtained from the combined observations than would be possible from the sum of the observations taken independently. All six satellites, Aqua, CloudSat, CALIPSO, PARASOL (French), Aura, and the Orbiting Carbon Observatory, will cross the equator within a few minutes of one another. Aqua was launched in 2002. There are a number of additional missions being developed by other nations or in partnership with the United States, and some, most notably Cryosat, have a specific polar mission. Coordination of satellite observations from this ever-growing international suite of sensors and additional focus by higher data rate sensors that do not collect data continuously would secure valuable benchmark datasets and advance the effort to assess the ongoing polar change.

Another IGY-era technology that has been advanced considerably since then is subsurface drilling, which provides information not accessible by satellite observation. Novel drilling technologies that could be adapted to do new science in the polar regions include directional drilling for retrieving ice cores and fast drilling for access of subglacial conditions. Directional drilling could retrieve ice core samples from locations at angles to the main vertical core and, at selected depths, generate multiple cores from a single surface platform. These additional samples allow for observations of the ways in which life forms have evolved and adapted to past climate and environmental change events. This technology forms the basis for new science of linked environmental and life system dynamics from ice cores. A fast-access mobile drilling system capability could be developed to rapidly drill arrays of boreholes through polar ice sheets 3 to 4 kilometers thick to allow targeted sampling from spatially distributed boreholes. Such a fast-access drill would enable a wide variety of research, from investigating conditions under ice sheets to borehole paleothermometry. Directional drilling and fast access drilling are employed in geophysical prospecting, but special design criteria need to be considered for drilling ice. There is also a need to develop drilling systems to be used for stratigraphic drilling to obtain paleoclimate records in the marine nearshore environment, from both fast sea ice and ice shelf platforms. Innovative technologies will allow continuous wireline coring, better core recovery, recovery of strata to a depth of 1,000 meters below the seafloor, and recovery of soft sediments. Other technologies must be developed for shallow water, such as shallow penetration drilling on polar continental shelves from ships.

TABLE 5.1 Some Future Satellites that May Be Useful During IPY 2007-2008

Satellite	Mission Objective
Aeronomy of Ice in the Mesosphere (AIM)	Explore why polar mesospheric clouds, also called noctilucent clouds, form and why they are changing
Aura	Study the chemistry and dynamics of the Earth's atmosphere by measuring ozone, aerosols, and several key atmospheric constituents that play an important role in atmospheric chemistry, air quality, and climate
Cloud-Aerosol Lidar and Infrared Pathfinder Satellite Observations (CALIPSO)	Provide new information about the effects of clouds and aerosols (airborne particles) on changes in the Earth's climate
CloudSAT	"Slice" through clouds to see their vertical structure, providing vertical profiles of cloud liquid water and ice water contents and related cloud physical and radiative properties
Landsat Data Continuity Mission (LDCM)	Extend the Landsat record of multispectral, 30-meter resolution, seasonal, global coverage of the Earth's land surface
National Polar-orbiting Operational Environmental Satellite System (NPOESS)	Monitor global environmental conditions related to weather, atmosphere, oceans, land, and near-space environment
NPOESS Preparatory Project (NPP)	Extend the measurement series being initiated with EOS Terra and Aqua prior to NPOESS
Orbiting Carbon Observatory (OCO)	Precise mapping of atmospheric carbon dioxide to enable more reliable forecasts of future changes in carbon dioxide and the possible effect on the Earth's climate
Solar-B	Determine solar origins of space weather and global change and the mechanisms for solar variability
Solar Dynamics Observatory (SDO)	Study active region formation on the Sun from subsurface to corona and monitor solar irradiance
Solar-Terrestrial Relations Observatory (STEREO)	Trace the flow of energy and matter from the Sun to the Earth and provide unique alerts for Earth-directed solar ejections
Space Technology 5 (ST5)	Miniaturized satellites will map the intensity and direction of magnetic fields in the inner magnetosphere
Time History of Events and Macroscale Interactions during Substorms (THEMIS)	Investigate magnetospheric substorm instability, a dominant mechanism of transport and explosive release of solar wind energy
Two Wide-angle Imaging Neutral-atom Spectrometers (TWINS)	Three-dimensional visualization and resolution of large scale structures and dynamics within the magnetosphere to understand the global aspects of the terrestrial magnetosphere

Sir Ernest Henry Shackleton's ship, the *Endurance,* was crushed by ice in 1914, but that is unlikely to happen to today's polar icebreakers, like the U.S. Coast Guard's *Polar Star,* shown here. The *Polar Star* and her sister ship the *Polar Sea* are both 122 meters long. There are 154 crew members per ship and each ship can accommodate as many as 20 scientists. The *Healy,* the newest icebreaker, is specifically designed to conduct research activities, providing more than 4,200 square feet of scientific laboratory space, numerous electronic sensor systems, oceanographic winches, and accommodations for up to 50 scientists. These vessels have sufficient hull strength to absorb high-powered rams into the ice. The curved bow allows the ship to ride up on top of the ice, then the bow is levered through the ice like a giant sledgehammer. The *Polar Star* is able to ram her way through ice up to 21 feet (6.4 meters) thick and steam continuously through 6 feet (1.8 meters) of ice at 3.5 miles per hour (5.6 kilometers per hour/3 knots). SOURCE: U.S. Coast Guard.

Access to submarines, icebreakers, and long-range research aircraft has proven valuable in the past, and increased access to them for the IPY would provide important new data that are difficult to gather in other ways. For instance, use of instrumented submarines in the Scientific Ice Expeditions program during the 1990s provided new insights into the nature of the Gakkel Ridge. Instrumented submarines could provide new insights into linked physical-chemical-biological systems during the next IPY. The polar icebreakers, currently in a state of declining capability, are needed to provide an important platform for access to interdisciplinary observations. A research aircraft with long-range flight capability, possibly an instrumented C-130, would provide monitoring of coastal and interior Antarctica and of the Arctic with spatial and temporal resolution not possible from satellites.

McMurdo Station (right) is the largest base in the Antarctic. The base can house some 1,000 people and support science of almost every discipline. Its modern laboratories and facilities are a far cry from Discovery Hut, November 4, 1911, on the left, built by the British explorer Robert Falcon Scott during his first Antarctic expedition. Scott later perished in a blizzard during his second expedition. SOURCES: Herbert Pointing. British Antarctic Expedition 1910-1913 (left) and George Somero, Stanford University (right).

Autonomous vehicles offer an entirely new tool for polar studies, and they can perform some functions that suborbital platforms, ships, and submarines cannot. For instance, there have been great advances in unmanned aerial vehicle (UAV) technology. The greatest benefits of using UAVs are that human lives are not endangered, they can fly in difficult weather conditions, they can remain aloft for long time periods (more than 24 hours), and they have tremendous range (Holland et al., 2001). Although still in their infancy, UAVs have been used in the polar regions to map sea ice conditions and atmospheric boundary layer conditions; the potential research applications are numerous, and their continued adaptation for polar operations would be greatly advanced by a concerted IPY research program.

Autonomous underwater vehicles (AUVs) are the subsurface complement to UAVs and share many of the same advantages—for example, long mission times and range with no risk to human life. AUVs have many applications for measurements under the floating ice shelves, the ice pack, and on missions in the open waters of the polar seas. They could be deployed for detailed exploration of subsurface features and for the study of hydrothermal systems and associated ecological systems. In addition, it would be possible for biologists to investigate questions of ecosystems and processes that, up to this point, have received little if any study. Another promising strategy for obtaining significant surface and near-surface observations is the deployment of instrumented robotic vehicles such as those in design and consideration for use on Mars. These vehicles travel slowly but are steadily powered by solar or wind energy and can accomplish a major traverse in a 100-day seasonal deployment. An international fleet of suitably designed rovers, each measuring useful variables relating to ice, snow, atmosphere, radiation, wildlife, chemistry, and so forth, hold extraordinary promise for collection of fundamental data across the most remote polar areas. In addition, the rovers could collaborate in such tasks as monitoring each other's activities, aiding in calibration and maintenance, and the like.

ASSEMBLING LONG-TERM MULTIDISCIPLINARY
POLAR OBSERVING SYSTEMS

A legacy of the IGY was the initialization of environmental monitoring records, such as the carbon dioxide record, which have proven key to documenting climate change and variability (NRC, 2004a). The intensive activity of the IPY 2007-2008 will extend measurements to include observations of linked physical, biological, and chemical observations of the atmosphere, oceans, ice, and land and will improve spatial and temporal coverage to provide a critical benchmark dataset for assessing the state of the polar environment. The infrastructure set in place will provide for long-term, spatially distributed, interdisciplinary observing networks to understand and explore the polar regions in the coming years and decades. The demand for observing systems that provide long-term measurements of important components of the Earth system comes from many fields. Climate change scientists (NRC, 2001b) recommend observing key state variables such as temperature, precipitation, humidity, pressure, clouds, sea ice and snow cover, sea level, sea surface temperature, carbon fluxes and soil moisture, and greenhouse gases. Oceanographers (NRC, 2004b) have recommended observing systems for key characteristics of deep-ocean features, including interdisciplinary measurements of physical, chemical, biological, and geological features. Biologists could utilize remote monitoring of organisms' body temperature, heart rate, blood pressure, and oxygen. Although the polar regions play key roles in Earth systems, there is a paucity of continuous measurements, with extremely poor spatial resolution. Recently developed sensors, communications technologies, and the ever-decreasing size and power restrictions of sensors all hold promise for developing a more extensive Arctic and Antarctic observing network.

The development and installation of international, long-term, multidisciplinary observing networks could be a particularly significant legacy of the IPY 2007-2008. These observing systems would provide scientists and decision makers with real-time information on the evolving state of the polar regions for decades to come. Stations that remain relatively fixed in place, such as on land or on stable ice sheets, as well as stations moving with the ice and seas, can be developed to integrate physical, biological, and chemical measurements. Innovative technologies could be applied so that sensors can vary their observational parameters or temporal sampling rates and interact with each other in a "sensor web." Sensors that can change the observation parameters or sampling rate when they sense unusual readings, such as a magnetic storm in Earth's upper atmosphere, are particularly valuable for the polar regions given their remoteness and the fact that in situ measurements are difficult to obtain. Some leading-edge sensors are even self-activating; that is, they remain dormant until the phenomenon they are programmed to measure occurs. The polar environment is an ideal proving ground for advancing these concepts, with tangible benefits from improving the relative proportion of valuable data that are collected to the more efficient use of available power. The next IPY offers the opportunity to realize permanent gains in this area by bringing together an international set of ideas and creative engineering approaches.

Innovative technologies now allow the design of polar observing systems that are less constrained by size, power, and data handling requirements than were previous measurements. For instance, miniaturization has been a hallmark of scientific instrument development since at least the beginning of the twentieth century, and over the past 50 years each decade has seen reductions in instrument size. This facilitates

Science has had a role in Barrow, Alaska, since an observatory was built during the first IPY in 1882-1883. The Naval Arctic Research Laboratory (NARL) was established in 1947, and its facilities were turned over to the Barrow Arctic Science Consortium in 1980. SOURCES: 1884 War Department Report prepared by expedition leader, P.H. Ray. U.S. Army Signal Corp. and U.S. DOD Navy photo, Office of Naval Research and Former NARL. Photo organized by Leslie Nakashima (top) and John Kelley, University of Alaska-Fairbanks (bottom).

deployment on smaller, more agile platforms. Power systems, their capacities, their endurance, and their survivability under harsh polar conditions continue to be important constraints for polar science and logistics. This is a problem faced at all scales, from the significant power consumptions of large year-round stations to providing a stable supply of power to the smallest microsensors. There have been many recent advances in fuel cell, solar, radioisotope thermoelectric generators, and other technologies that can be marshaled for IPY 2007-2008 and adapted to the appropriate scale. As the efficiency and reliability of autonomous vehicles and stations increase, and until the communications bandwidth and availability also increase, the need for better data storage capabilities will continue to rise sharply. This poses challenges for engineers and computational scientists who must be engaged to work with polar researchers in resolving the design issues associated with increased data memory and storage capacity in polar environments.

Increased base infrastructure to support people doing research in the polar regions is needed to advance understanding and discovery. Remote systems require servicing and support, and collected data and samples need security and forwarding. Along with remote systems, human-intensive field operations require improved research base infrastructure and reliable logistical support. These needs are common to both polar regions. In the American Arctic, for example, the former Naval Arctic Research Laboratory facility outside Barrow has offered researchers space and logistical support from renovated but formerly derelict Quonset hut buildings that were constructed in the 1940s (Norton, 2001). The Barrow-area situation is being improved by construction of the 20-laboratory 50-bed Barrow Global Climate Change Research Facility. This is the first step in addressing modern support requirements of an increasingly expanding and interdisciplinary Arctic research program based on Alaska's North Slope. The construction of sustainable research facilities at both polar regions must be coupled with significant data capture and transfer capacity and remote support capabilities in order to meet today's and currently envisioned future research requirements.

Research bases are not the only infrastructure in need of upgrading. Icebreakers have fallen into disrepair; logistical support, including helicopters, is exceedingly scarce; submarine capability needs to be updated and resources allotted to the support of basic research falls at least an order of magnitude below that available elsewhere. Advances in aeronautics and navigation have steadily increased access and safety of flight operations in the polar regions, and some key Arctic stations now have year-round access by air. However, year-round access to most of Antarctica, excepting the Antarctic Peninsula, is not routine (winter flights mostly are for humanitarian reasons). Although winter flights to high-altitude Antarctic ice sheet stations continue to be risky, extending access to other regions may be feasible and would open new scientific opportunities even if flight operations were extended to encompass the nine weeks or so after sunrise and before sunset.[1] The development of the Antarctic Mesoscale Prediction System provides the weather forecast tools to enable expanded flight operations anywhere in Antarctica.

Finally, enthusiasm to deploy innovative sensors and instruments must be balanced with potential environmental impacts. For many decades, scientists and engineers have

---

[1]McMurdo sunrise is on August 19 and the sunset occurs on April 24. There is some sunlight and twilight each day over the nine weeks that were studied as a candidate for extended flight operations.

concerned themselves with the development of sterile instruments for the exploration of new environments such as the planets and their moons. The Antarctic Treaty and related conventions and protocols set rigorous environmental standards for science. Whenever probes are sent into previously unexplored environments in the polar regions, they may be deployed only after their environmental efficacy has been demonstrated. Environmental risks in the exploration of unexplored polar environment should not be taken until the efficacy of instrumentation is demonstrated. Using the impetus of the IPY 2007-2008 to develop sound, proven instrumentation may delay some observations, yet development during the IPY will provide a basis for exploration and observation in the decades to come.

INNOVATIVE DATA ANALYSIS TECHNOLOGY

The international cooperative research programs envisioned for the new IPY will produce large amounts of data that ultimately will need to be analyzed during and after 2007-2008. A few of the innovative types of data analysis likely to contribute to IPY science are highlighted here.

The rapid advances in computing technology since the IGY have provided scientists with tools to perform statistical analyses and develop dynamical models that were not previously possible. Statistical analyses can uncover spatial patterns and temporal trends in data, and they can be used to identify environmental relationships and test theoretical models (Zwiers and von Storch, 2004). The availability of powerful computers underlies new modeling efforts, and the IPY could advance U.S. effectiveness in climate modeling by applying the recommendations of a previous National Research Council report on improving climate modeling (NRC, 2001a).

Genomic and bioinformatic data analysis will play an important role in diverse types of biological research in the polar regions. The key roles that genomic techniques will play in polar biology were indicated in Chapter 4. However, data gathering is only the first step in exploiting the potentials of genomics. High-throughput DNA sequencing generates enormous datasets, which must be annotated and stored in readily accessible formats to ensure broad use (NRC, 2003a). Analysis of gene expression, using the techniques of proteomics (characterizing the suite of proteins present in a sample), likewise involves massive datasets and creates challenges for making the data easily interpretable and accessible to a broad range of users. Standardization of techniques for data gathering and analysis through bioinformatic methods will be necessary to allow the full potential of contemporary genomic science to contribute to polar biology.

Data assimilation focuses on the assessment, combination, and synthesis of observations from disparate sources to reproduce the evolution of a system and forecast its behavior. For instance, merging observations with short-term atmospheric forecast model output is a technique for optimizing the impact of the spatially sparse and temporally limited observations typically available in the polar regions. Data assimilation for high latitudes, which is the basis for atmospheric reanalyses, faces challenges due to imperfect understanding of physics, omitted processes, numerical limitations, and diverse data sources, especially from space. These deficiencies need to be overcome before high-quality reanalyses can be produced for the period from the IGY to the present. Further, similar improvements need to be extended to the land surface and ocean to describe the variability and trends of the fully coupled polar environment of the past and into the future. Such an accomplishment would result in a useful extended view of recent polar climate. An additional result of improved coupled models would be the very tangible advantage of more reliable forecast models to support air operations, especially in the Antarctic.

DATA ARCHIVING AND DISSEMINATION

The efficient handling, storage, and dissemination of data will be paramount for a successful IPY effort. Datasets, and ancillary information such as metadata, must be preserved for decades and stored in ways that promote access. It also will be critical to facilitate the integration of multiple types of data, and to extract the full scientific and societal value of IPY research. The data must be available in appropriate formats for scientists, public- and private-sector decision makers, and managers. During IGY, about 50 permanent (physical) observatories were set up in the Arctic and Antarctic, and the World Data Center System was established to ensure that the data collected were properly archived and made available without restrictions for scientific research and practical applications. The IPY 2007-2008 may expand and augment the current system or even establish a system of virtual data centers. This concept implies free access to all available data through the internet; the proposed Electronic Geophysical Year (Baker et al., 2004) would focus specifically on this concept. However, large repositories of historical data, some from the IGY, still reside on hard copies or magnetic tapes and are inaccessible for most scientists and at risk of permanent loss without concerted rescue efforts. The IPY should also focus on extending the data record back in time.

Data sharing and, as noted in the next chapter, education and outreach will be important components of the next IPY. Thanks to satellite communications, transmission of large quantities of data from the polar regions to research laboratories has become routine. The IPY could also encourage utilizing new communications technologies to reach out and interact with U.S. students and citizens from the reaches of the polar regions. One unresolved communications issue for the polar regions is that the orbital configuration of communication satellites leads to "blackouts" in data transmission at certain times. IPY 2007-2008 planners and U.S. department and agency leaders could use the IPY to inventory present and anticipated data transmission needs and determine whether it is feasible to reposition some satellites to provide more continuous coverage for stations in the high northern and southern latitudes. Society is moving rapidly to broadband communications for data dissemination. Several U.S. federal agencies, including the National Science Foundation, NASA, and the Energy and Defense Departments have supported the creation of the infrastructure, including the primary backbone linkages and the computation, visualization, and memory capabilities that have made today's scientific transformation possible. While high-speed broadband communication has begun to link scientists in most industrial nations, extension of high-speed broadband networks to stations in the polar regions remains limited. Both the science community's need for real-time data and the education and outreach community's need for active interaction with the polar community drive the need for expansion of broadband communication.

Impacts of Environmental Change on Ecosystems

When Berry Lyons steps off a plane and walks out onto the McMurdo Dry Valleys in Antarctica, he's greeted with as desolate a landscape as any he is likely to find on Earth. Instead of an immense sweeping expanse of ice and snow—like 98 percent of the continent—the Dry Valleys are deserts, receiving less than 4 inches (10 centimeters) of water each year.

"The stark beauty of these valleys is breathtaking," says Lyons, director, Byrd Polar Research Center and professor in the Department of Geological Sciences, Ohio State University—who has been going to the Antarctic since 1981. "It is difficult to believe that life exists in these soils, ephemeral streams, and perennially ice-covered lakes—but it does. Because of this, the Dry Valleys have been used as analogs to extraterrestrial landscapes since the 1960s."

As a geochemist, Lyons is less curious about extraterrestrial similarities as he is about whatever clues the chemistry of streams, lakes, and glaciers may reveal about the interconnectedness between biology and climate. He and his colleagues in the McMurdo Dry Valleys Long-Term Ecological Research program are investigating the role of climate on this extreme ecosystem. Essentially, Lyon's team is collecting stream, ice, and lake samples—as well as stream sediments and some soils—and analyzing these to see who's there and what else the chemistry may reveal.

"These valleys are extremely sensitive to small changes in climate," he says. "For this reason, we are assessing the impact of present and past climatic change on the structure and function of this polar desert ecosystem. While these variations in climate are small by temperate standards, they have great ecological consequences here in the Dry Valleys."

According to Lyons, the key climatic parameters influencing ecosystem structure and function in the McMurdo Dry Valleys are the ones that affect the physical state of water. Temperatures rise to a few degrees above freezing in late December and January, producing glacial melt water that, in turn, exerts the primary influence on the Dry Valleys by replenishing water and nutrients to ecosystems there.

"Variations in temperature, snowfall, and solar radiation influence melting water production on the glaciers, and the resulting stream flow influences stream microbial populations," he says. "The closed basin lakes that are fed by the streams can respond in a variety of ways. Phytoplankton populations, for example, at the bottom of the food chain, may change. On-going studies are looking at life on the glaciers, in the streams, lakes, and soils."

Understanding the interrelationships between the dynamics of ecological systems and climate is important in predicting how the ecosystem will respond as climate changes, Lyons says. "The International Polar Year 2007-2008 will allow us to develop a better understanding of how polar systems are linked to those in other parts of the Earth." For example, what are the factors that control biodiversity in the Antarctic terrestrial and limnetic ecosystems, and how does the diversity vary on a continental scale? There is a need to bring scientists working on terrestrial and lake ecology from all nations together to collect and exchange data, so a better perception of these very climate sensitive systems can be developed.

"Hopefully, the International Polar Year will help lead to a more comprehensive view of Antarctic ecology and its response to climate."

Principal investigators in the McMurdo Dry Valleys Long-Term Ecological Research Program taking a break on a glacier: (Left to right in back row) John Priscu, Montana State University; Berry Lyons, Ohio State University; Diane McKnight, University of Colorado; and Diana Wall, Colorado State University. (Left to right kneeling) Ross Virginia, Dartmouth College; and Andrew Fountain, Portland State University. SOURCE: Diana Wall.

6

Increasing Public Understanding and Participation in Polar Science through the International Polar Year

"Words are things; and a small drop of ink,
Falling like dew upon a thought, produces
That which makes thousands, perhaps millions think."

LORD BYRON

The public's interest in polar regions is profound and deep rooted. The popularity of recent films and books on Shackleton's journey, the prevalence of polar images in everyday culture, and the popularity of adventure literature all exhibit an inherent interest in the region, especially when it is linked with exploration, discovery, adventure, isolation, self-reliance, hardship, passion, history, and exotic landscapes and biota. The International Polar Year (IPY) is an excellent opportunity to build on this inherent interest and engage the public in science targeting all ages through the use of different media. Developing education and outreach programs for the IPY 2007-2008 will take a concentrated effort by formal and informal educators and the media working closely with active scientists. As with the International Geophysical Year (IGY), this public outreach holds the potential to leave an enduring legacy by fostering a new generation of scientists, engineers, and leaders. The education and outreach strategy for the IPY 2007-2008 will develop a diverse range of opportunities with actual participation by students of all ages.

Embracing the possibilities for interactive programs made possible by emerging technologies, IPY 2007-2008 education, training, and outreach should be extended to all age groups while building on successful existing models. Fortunately, we live in an age of extraordinary communications, especially compared to previous IPYs and the IGY; new approaches should be taken that are interactive, that make use of diverse media, and that provide opportunities for hands-on participation by the public. Education and outreach efforts could target the next generation of scientists by including opportunities at both polar regions, targeting underrepresented groups and minorities, and promoting international understanding.

65

Few animals can survive in Antarctica's cold, dry interior, but many thrive along the shore and in surrounding waters. Among the animals that make their home in the Antarctic are penguins, which can dive almost 900 feet (275 meters)—about the length of three football fields—for a meal. Some species can hold their breath for 20 minutes. SOURCE: National Oceanic and Atmospheric Administration.

Building a strong and lasting bridge between the polar regions of the Earth and the polar science community with the public and students of the United States will require innovative integration of methods, media, and forums enabled by modern communication and data technologies. Elegant posters for classrooms, the *Scholastic Weekly Reader*, and short film clips were the principal approach of the IGY in 1957-1958. For the IPY 2007-2008, we must build on the diverse array of educational methods and media outlets now available so that the full range of learners—from retired citizens in Arizona to elementary school children in Maine—are engaged in polar assessment and exploration. Some fundamental building blocks of the IPY 2007-2008 education and outreach programs will be:

• Field experiences for teachers and students of all ages who can serve as ambassadors to the diverse educational and media outlets and who may become the future scientific leaders.
• Remote participatory experiences enabled by high-bandwidth communications and the wealth of existing materials on polar science for students and learners of all ages, touching the wired classrooms of the nation and the active and strong national network of science museums.
• Leveraging the network of polar-savvy reporters, authors, artists, and filmmakers who have been involved in study of the polar regions and who remain a tremendous resource.

HANDS-ON PARTICIPATION IN THE IPY

The wonders of the polar regions are best experienced through active participation. Scientists routinely take in situ observations in the polar regions, but the IPY might expand field participation to include all education levels and ages. At the most basic level, the IPY endeavor could train future scientists, engineers, and leaders by placing a special emphasis on fieldwork during 2007-2008. Formal programs, such as the Research Experience for Undergraduates, already exist to provide field experience to undergraduates, but there are few formal programs that can reach out to a large percentage of undergraduates and graduates who are interested in polar field experience.

In K-12 classrooms, teachers have a tremendous role in communicating the science of change and exploration to young students. Field experiences for adventurous teachers, coupled with programs that can be easily implemented at the national network of science museums, are needed to help expose the polar regions to today's youth, who are tomorrow's leaders. Another untapped resource for entraining students in polar studies is through the Boy Scouts of America and the Girl Scouts of America. For instance, Paul Siple went to Antarctica as a Boy Scout on the first Byrd Expedition (Siple, 1931, 1959) and Eagle Scout Dick Chappell went to Antarctica during the IGY. Another novel product that could result from IPY research is the creation of children's books and outreach to scholastic publishers to include more IPY research in textbooks.

For the public, ecotourism can promote increased public participation and understanding of the polar regions. Tourist ships could participate in the science of IPY by

Teacher Daniel Solie explains earthquakes and wave motion to children and adults at Gakona School in Gakona, Alaska, November 21, 2002. The IPY will offer a range of educational opportunities both in the classroom and in the field. SOURCE: Suzanne McCarthy, Prince William Sound Community College.

taking in situ measurements, and expert lecturers on cruises, along with field trips taken off ship, could make indelible impressions on participants. There is some concern that ecotourism is a double-edged sword in that the growing number of cruise ships and tourists may negatively impact the largely undisturbed polar environs. However, tour operators in the Antarctic region have been working with the various governments involved to minimize their footprint.

The tangible benefits of experiencing the polar regions are easily shown through reflection on the IGY. A cadre of people parlayed experience in Antarctica into graduate programs, and several IGY-era people, such as Kirby Hanson and Mario Gioveinetto, went to the South Pole without a bachelor's degree and have gone on to obtain doctorates and do significant polar research. Moreover, many participants in IGY who visited Antarctica in 1957 and 1958 returned to school to get Ph.D.s, and many became professors or researchers in nonpolar fields, highlighting the fact that many other fields and society as a whole benefited from the field experience of these individuals.

GLOBAL PARTICIPATION IN THE IPY

Although field courses can bring students, teachers, and the public to the polar regions, novel education programs, the proliferation of communication technologies, and the internet can bring the polar regions to classrooms and homes and reach out to people who cannot or do not want to experience the polar regions firsthand. The continued and growing emphasis on education by agencies, consortiums, and professional societies could serve as the basis for virtual participation in the IPY, educating the next generation of scientists, engineers, and leaders. For instance, many of the large granting bodies, such as the National Science Foundation, the National Aeronautics and Space Administration, the National Oceanic and Atmospheric Administration, the Department of Defense, and the Department of Energy, have offices for education, and for the next IPY a strong interagency effort for education must begin well in advance of 2007-2008. Many nonprofit organizations, foundations, and university consortia, such as the University Corporation for Atmospheric Research and the University of the Arctic (see Box 6.1), can also advance the goals and themes of the IPY.

BOX 6.1
The University of the Arctic

The University of the Arctic (UArctic) is an international non-governmental organization dedicated to higher education in and about the circumpolar north. UArctic partners in the United States include the University of Alaska-Fairbanks and Illisagvik College in Barrow. UArctic is a decentralized university without walls that mounts programs of higher education and research, builds local educational capacity, and stimulates cooperation among participating institutions. Courses in the Bachelor of Circumpolar Studies are available both online and locally at member institutions. Students may also participate in exchange programs with another UArctic partner. Students, researchers, and instructors can participate in field courses across the Arctic. The Northern Research Forum and Circumpolar Universities Association organize major conferences on northern issues and provide opportunities for the academic and scientific community to interact with other northern stakeholders.

Professional societies also may share the IPY goal for education and outreach. More than a dozen societies have a polar mission, and many have their own educational outreach programs, including grants-in-aid for research and travel assistance for participation in conferences as well as magazines and journals. The societies' network of activities could be encouraged to promote IPY projects. In addition, significant industrial development now exists in the Arctic. Through their philanthropic programs, these companies may be interested in sponsoring certain IPY education activities. By linking individual philanthropists with university initiatives and activities, new opportunities for students can be created.

An increased appreciation for the polar regions can also be achieved through more informal means, such as exploratoriums, museums, and competitions like the Odyssey of the Mind (see Box 6.2). Building on successful programs run by groups such as the San Francisco Exploritorium could help to reach hundreds of thousands of students through live feeds from the polar regions. Expanding this program nationally and linking it with teacher training would provide real participation across the country. Public museums, such as the Smithsonian Institution, attract many visitors each year and offer an excellent virtual outreach opportunity through interactive displays, Web cams, and more traditional museum pieces.

Perhaps the greatest opportunity for interaction lies with the internet. It is the modern tool that can provide virtual experience, enabling the public to become involved in the IPY. Interactive websites, streaming video, and distributed computing all offer opportunities for participation in the IPY. The next-generation advancement of the internet (http://www.internet2.edu) will make it even more feasible to create interactive websites and will facilitate effective and rapid transfer of IPY streaming video. Distributed computing parses small parts of a problem to many computers and then combines the results into a solution. Recent distributed computing projects to look for extraterrestrial radio signals and to find more effective anti-AIDS drugs used hundreds of thousands of volunteer computers all over the world via the internet. Websites currently exist to link interested volunteers with projects (e.g., http://www.aspenleaf.com/distributed/), and the IPY might use distributed computing for a variety of uses, including polar climate modeling and genomics.

BOX 6.2
The Odyssey of the Mind

The Odyssey of the Mind School program fosters creative thinking and problem-solving skills among participating students from kindergarten through college. It features an annual competition component at local through international levels. Students solve problems in a variety of areas, from building mechanical devices such as spring-driven vehicles to giving their own interpretation of literary classics. Through problem solving, students learn lifelong skills, such as working with others as a team, evaluating ideas, making decisions, and creating solutions while also developing self-confidence from their experiences. More information on the Odyssey of the Mind is available at http://www.odysseyofthemind.com/.

CREATIVE MEDIA APPROACHES

In addition to direct and virtual participation in IPY activities, the IPY 2007-2008 will be more successful if the public is given information about how the study of polar processes is of importance to daily life and to global processes in general. This effort will be facilitated by translating research activities and results into everyday knowledge by innovative use of the media, thus giving the public the chance to be involved in the IPY. Public outreach for the IPY likely will involve a mix of traditional technologies, such as radio and print media, with recent advancements, such as satellite TV and the internet.

The major news networks and the nation's prominent newspapers collectively remain the largest and most effective portal to reach the public. Embedded journalists from these sources can be an information portal for the public, and recent successful examples, such as *New York Times* reporter Andrew Revkin's trip to the North Pole Observatory, highlight the tremendous public interest and potential for embedding journalists.

Popular science publications and national newspapers have always been important in bringing polar activities and research to U.S. citizens, while educational programming on nature and history is in such demand now that there are entire television networks devoted to each. Traditional nature shows such as "Nova" or TV series documenting polar history and culture would highlight the IPY, but additional programming directed at younger viewers may have broader appeal. Documentaries about young Inuit life, an Eagle Scout's trip to the ice, or a graduate student's first season in the Arctic or Antarctic may inspire other young people to explore the polar regions further, either through a career in science or as a lifelong interest.

Specialty networks, such as the NASA Channel, The Science Channel, and the National Geographic Channel, also can provide in-depth coverage of IPY activities throughout the IPY-intensive field campaign. Live video expeditions or similar real-time polar educational programs also could stimulate public attention by profiling prominent polar research expeditions during the IPY. On the radio, programs including "Radio Expeditions," a collaboration between National Public Radio (NPR) and the National Geographic Society, could produce feature stories on the IPY for NPR's *Morning Edition.* The general themes for stories are the natural world and threatened environments, diverse cultures, adventure, and exploration and discovery (http://www.npr.org/programs/re/index.html).

The key to securing coverage of IPY on any of these outlets is an effective public information arm. The template for productive relations with the media is the National Science Foundation's (NSF) Office of Polar Programs. The benefit is that NSF is able to bring public and political attention to the research programs it supports, informing both taxpayers and elected officials of the value and achievements these research efforts provide. Any public outreach plan will, of course, rely on the existing public information offices of the various universities and agencies whose researchers are involved in IPY.

Lessons from the Otters

In the turquoise blue water of Simpson Bay, Prince William Sound, Alaska, an otter floats on her back, basking in the summer sun. In the twinkling of an eye, her puppy is born. The new mom scoops the puppy gently from the water and places him on her stomach where she grooms him as he snuggles in to nurse. Some time later, the mom leaves her buoyant puppy floating on the water while she dives to search for food.

In a small boat, about 100 feet (30 meters) away, biologist Randall Davis and two volunteers from an Earthwatch research expedition are taking digital images of the female for photo-identification and noting her location and behavior for a long-term study about sea otters. A professor of Marine Biology at Texas A&M University, Davis is interested in the cellular, physiological, and behavioral adaptations of marine birds and mammals—including otters. He has been working with marine animals in the Arctic and Antarctic since 1976, and in 1989 directed Exxon's Sea Otter Capture and Rehabilitation Program after the disastrous oil spill in Prince William Sound.

"In a broad sense, this research will improve our understanding of sea otter ecology and the factors that influence fluctuations in the population," says Davis. "Results from this research will have their most direct application to the dramatic decline (about 90 percent) in the sea otter population in parts of the Aleutian Islands and southwest Alaska. The U.S. Fish and Wildlife Service and the U.S. Department of the Interior are seeking information that will provide insight into the cause of the decline. Sound policies based on scientific data are essential to manage and, if possible, mitigate the current decline in Alaskan sea otters."

While most of his research over the years has been conducted as a university researcher, this changed in 2001 when Davis started leading expeditions north each summer. In a specially developed outreach program, Earthwatch Institute engages people worldwide in scientific field research and education to promote the understanding and action necessary for a sustainable environment.

SOURCE: Randall Davis, Texas A&M University.

continued

Simpson Bay is a good location to work because it is teeming with wildlife, Davis says. "About 120 otters make their home in the bay, where there is a good cross-section of otters, and where about 20 to 40 puppies are born each summer," he adds. "It seems as though these pups are born in something like 10 or 15 seconds. It happens fast. You can be looking at a female, turn away and look back and, suddenly, she's got a puppy!

"We are doing some very basic research on sea otter ecology," he explains. "First, we photograph the otter's face using a digital camera and a telephoto lens, so we can distinguish one individual from another. We can identify individuals often because many otters—especially females—have pink scars on their black noses, which result from mating behavior. In recognizing distinct individuals, we can begin to ask questions about individual behavior and movement patterns. We can actually follow those individuals across time."

"We also collect some basic information on this individual. If the otter is feeding, for example, we may stay for a while and record dive depth, duration, location, and see what it brings to the surface. If we find a territorial male, we may be looking for different types of behavior such as patrolling, interacting with females, or mating. We follow these males around and map out their territories. We've also been characterizing the physical environment so we can better understand why otters are using certain parts of their habitat and not others."

Davis says that his volunteers are highly motivated individuals, age 18 to 80, who are very interested in nature. Over the summer, he works with seven teams of six persons each. "Each team comes in for about 10 days and then we have four days between teams, so we have a two-week cycle. These six-person teams are then broken up into two teams of three. Usually, each three-member team, plus me or one of the other research leaders, goes out every morning or afternoon on small boats to make observations and collect data.

"Being out there in the wilderness with, not just sea otters, but with a whole range of animals—from bears to mink to seals to seabirds—makes a lasting impression on these individuals," says Davis. "It certainly gets them thinking not only about biology, but about conservation. We have fairly long discussions in the evening about all these types of issues." Not surprisingly, a number of volunteers include teachers who join the expeditions not only for their own enjoyment, but to go back and share what they have learned with their students—further multiplying the outreach endeavor.

Recognizing the need for conservation at the poles is one of the take-away messages of these expeditions and Davis sees this same potential for the International Polar Year. It is an opportunity to focus public attention and imagination on the very unique environment that exists at both the poles, on the exciting science that is going on, and on the importance of this research to citizens of lower latitude countries. "I think it will get people interested and excited, and that translates into additional support. It also raises awareness and support for those treaties and policies that protect these areas and the wildlife that is there."

7

Actions Needed to Make the International Polar Year Succeed

The International Geophysical Year (IGY) brought about fundamental and long-lasting changes in technology, international scientific cooperation, science, and human understanding of nature and the cosmos. The legacy of the IGY is ongoing, informing us about our planet and how it works. Two important international treaties, addressing issues for space and Antarctica, have their origins in the IGY.

Compared to 50 years ago, the most dramatic changes in how science will be conducted in the IPY are the temporal and spatial scales that can be addressed; the advanced sensors, methodologies, and technologies of the computer and information age that can be brought to bear; and the data access, archival, and public involvement that is assured by modern communications and information technologies. Breakthroughs and insights will follow as the new data lead to a deeper and better understanding of the polar regions and their role in the Earth system.

The International Polar Year (IPY) 2007-2008 will build on the legacy of the previous IPYs and the IGY. The IPY aspires to make the following lasting impacts:

1. Increase international partnerships and cooperation. The scale, scope, and spirit of the IPY demand international coordination and cooperation.

2. Elevate polar science to a new level. The IPY will be a rallying call to "think outside the box" and use interdisciplinary science to investigate new realms. The IPY will address issues that are most relevant to today's society that can be best addressed by study in the polar regions. IPY exploration and research will be of the highest quality and challenge the state of the art.

3. Launch and accelerate initiatives that would otherwise be slow in emerging. The most advanced technologies can be brought to bear in an ensemble approach. A spectrum of temporal and spatial scales will be addressed. Information will be accessible, preserved, and made available to all in the scientific and engineering communities.

4. Engage the public in polar discovery. The IPY is an unprecedented opportunity to engage the public in discovery on a real-time basis. Public outreach, education, and public engagement will be hallmarks of the IPY.

5. Establish human and technological infrastructure that will last far beyond the IPY years. Synergy and focus in science will create unpredictable and unforeseen outcomes greater than the sum of the parts. Long-term contributions will mature and evolve over the years as infrastructure is used, databases grow, samples are analyzed, and knowledge is disseminated or harvested.

RECOMMENDATIONS

Events observed today in the polar regions have captured the attention of the American public. The polar regions play a critical role in controlling linked global-scale atmospheric, oceanic, terrestrial, and biological processes; changes first seen in the polar regions affect weather and life in many areas, including weather patterns here in the midlatitudes. The changes we are now witnessing in the Arctic and Antarctic are unlike any in recorded history, yet we do not understand how or why they are occurring, and we lack the tools and knowledge to understand whether they are part of natural variability or due to anthropogenic effects (IPCC, 2001; NRC, 2002). We do not understand abrupt changes in the past, we cannot forecast the coming years, and we are unprepared to mitigate or adapt to the possible outcome. Life on all levels, from microbial to human, is affected by the environment, yet there are many unknowns in understanding adaptation potential and ecosystem dynamics. Virtually all of the phenomena studied in the polar regions, including biology, geophysics, oceanography, planetary physics, glaciology, economics, and sociology, have close linkages to worldwide processes. Thus, important societal issues, involving countries of all latitudes, are intrinsically linked to processes occurring in the polar regions.

The U.S. National Committee for the International Polar Year recommends that the following opportunities for important new science should receive attention starting in the International Polar Year 2007-2008.

Recommendation 1: The U.S. science community and agencies should use the International Polar Year to initiate a sustained effort aimed at assessing large-scale environmental change and variability in the polar regions.

• *Provide a comprehensive assessment of polar environmental changes through studies of the past environment and the creation of baseline datasets and long-term measurements for future investigations.*

Environmental changes currently observed in the polar regions are unprecedented in times of modern observation. Studies investigating natural environmental variability, human influence on our planet, and global teleconnections will help in understanding mechanisms of rapid climate change and in developing models suitable for forecasting changes that will occur in the twenty-first century. This effort will need to be sustained after IPY 2007-2008.

• *Encourage interdisciplinary studies and the development of models that integrate geophysical, ecological, social science, and economic data, especially investigations of the prediction and consequences of rapid change.*

Because of its broad interdisciplinary approach, research initiated in IPY 2007-2008 stands to make a significant contribution to our understanding of the causes and consequences of change in the polar regions.

Recommendation 2: The U.S. science community and agencies should pioneer new polar studies of coupled human-natural systems that are critical to U.S. societal, economic, and strategic interests.

• *Encourage research to understand the role of the polar regions in globally linked systems and the impacts of environmental change on society.*

Daily life and economic and strategic activities are constantly affected by changing environmental conditions, including the frequency and degree of severe weather events such as storms or droughts in many regions, including the continental United States. Investigations of impacts of linked environmental-technological-social change and health effects in many communities, including northern communities, are needed.

• *Investigate physical-chemical-biological interactions in natural systems in a global system context.*

Interdisciplinary approaches hold great promise for understanding the dynamics of anthropogenic activities, technologies, and environmental consequences. Investigations of linked atmospheric-oceanic-ice-land processes in the polar regions will enable understanding of global linkages and transformations due to natural and anthropogenic causes.

• *Examine the effects of polar environmental change on the human-built environment.*

Because of the recent large-scale environmental changes, northern communities, infrastructure, and other forms of human-built environment are affected by a variety of factors, such as the thawing of permafrost, higher frequency of severe storms and weather conditions; increased shore- and beach erosion, vegetation die-off, and fire danger. New engineering and policy research should investigate economically feasible and culturally appropriate mitigation techniques for countering the effects of a changing environment on technology, local communities, and their infrastructure, including all-season ground and air transportation, the design of roads, harbors, foundations, and buildings.

Recommendation 3: The U.S. International Polar Year effort should explore new scientific frontiers from the molecular to the planetary scale.

• *Conduct a range of activities such as multidisciplinary studies of terrestrial and aquatic biological communities; oceanographic processes, including seafloor environments; subglacial environments and unexplored subglacial lakes; the Earth's deep interior; and Sun-Earth connections.*

Opportunities for discoveries exist in many areas, and research could elucidate the structures of poorly understood biological communities, notably the microbial populations that contribute to most biogeochemical transformations; reveal oceanic processes that contribute importantly to biological productivity and climate; and discover new physical, chemical, and, potentially, biological characteristics of subglacial lakes long isolated from atmospheric contact. This research also could help understand major geological processes such as seafloor spreading, explore the subglacial topography and bedrock geology of regions important for Earth's climate history, map the structure of Earth's interior and explore the links between mantle structure and surface processes, and provide an integrative synthesis of the interactions of our planet with the Sun.

- *Apply new knowledge gained from exploration to questions of societal importance.*
Polar biological studies, notably those that employ modern genomic methodologies, will advance biomedical and biotechnological research. For example, understanding how small mammals withstand temperatures near freezing during hibernation will contribute to improved protocols for cold storage of biological materials and for cryosurgery. Studies of oceanographic phenomena will facilitate more accurate understanding of the mechanisms driving climate change. Understanding how increased flow of fresh water into the polar oceans alters circulation patterns and transfer of heat from the tropics to the poles is one example of contributions from oceanography. Advances in the geosciences (e.g., through study of the extremely slow seafloor spreading rates in the Arctic) may shed light on tectonic processes that contribute to seismic events. Better understanding of solar influences on the atmosphere and Earth will improve understanding of the forces that drive weather systems and of solar activity on global communications and other technical systems.

- *Invest in new capabilities essential to support interdisciplinary exploration at the poles.*
New scientific discoveries are based in part on the availability of enhanced logistics to provide access to unexplored regions as well as new technologies to provide new types of data. The IPY field component should aggressively seek to further develop innovative strategies for polar exploration.

Recommendation 4: The International Polar Year should be used as an opportunity to design and implement multidisciplinary polar observing networks that will provide a long-term perspective.

- *Design and establish integrated multidisciplinary observing networks that employ new sensing technologies and data assimilation techniques to quantify spatial and temporal change in the polar regions.*
The IPY will provide the integrative basis for advancing system-scale long-term observational capabilities across disciplines. A goal of the IPY should be the design and establishment of a system of integrated multidisciplinary observing networks. New autonomous instrumentation requires development with the harsh polar environment in mind. Instruments required for different types of studies can be clustered together, minimizing the collective environmental risks of survival and encouraging integrated analysis. Common observational protocols, such as observation frequency and measurement precision, will increase the spatial range of the observations and simplify data assimilation. Once established in the IPY, such protocols will serve polar science in the longer term.

- *Conduct an internationally coordinated "snapshot" of the polar regions using all available satellite sensors.*

Two hallmarks of the IGY were the dawn of the satellite era and the establishment of enduring benchmark datasets. Today's ever-growing suite of satellite sensors provides unique views of the polar regions with unprecedented detail. Marshaling the collective satellite resources of all space agencies around the world would supply generations of future scientists an unparalleled view of the state of the polar regions during the IPY 2007-2008.

Recommendation 5: The United States should invest in critical infrastructure (both physical and human) and technology to guarantee that the International Polar Year 2007-2008 leaves enduring benefits for the nation and for the residents of northern regions.

- *Ensure the long-term availability of assets necessary to support science in the polar regions, such as ice-capable ships, icebreakers, submarines, and manned and unmanned long-range aircraft.*

Although IPY 2007-2008 is planned as a focused burst of activity with demonstrable results, it should also provide long-term value and leave a legacy of infrastructure and technology that serves a wide range of scientific studies for decades to come.

- *Encourage development of innovative technologies to expand the suite of polar instruments and equipment, such as unmanned aerial vehicles (UAVs), autonomous underwater vehicles (AUVs), and rovers.*

Observational systems for the polar regions can be improved enormously by applying innovative technologies. Recent technological advances in UAVs, AUVs, and robotic rovers can be marshaled and adapted for the IPY to ensure that these platforms enhance IPY research capabilities.

- *Develop advanced communications systems with increased bandwidth and accessibility capable of operating in polar field conditions.*

The innovative technologies and large-scale field operations during IPY 2007-2008 will require advanced communications systems with high-speed, real-time access to communicate and distribute data from both polar regions to the rest of the world.

- *Develop international standards, policies, and procedures that ensure data are easily accessible for the current generation and permanently preserved for future generations.*

The data management systems should provide free and open access to data in standard formats. In addition, extensive metadata should be included to facilitate long-term reanalysis and so that datasets can be used by a variety of users. This effort should include data rescue efforts to expand the data record back in time and ensure that historical data are not lost.

- *Develop the next generation of scientists, engineers, and leaders and include underrepresented groups and minorities.*

Tomorrow's leaders are in today's classrooms, and the IPY effort should focus on cultivating an interest in the next generation of scientists, engineers, and leaders to create a lasting legacy.

Recommendation 6: The U.S. International Polar Year effort should excite and engage the public, with the goals of increasing understanding of the importance of polar regions in the global system and, at the same time, advancing general science literacy in the nation.

• *Develop programs in education and outreach that build on the inherent public interest of the polar regions and provide a broad lay audience with a deeper understanding of the polar regions.*
The polar regions have important direct and indirect effects on the rest of the world, and the IPY can help explain the importance of the polar regions to the public.

• *Create opportunities for education, training, and outreach for all age groups and build on successful existing models. Education and outreach during the IPY should include innovative new approaches that are interactive, make use of diverse media, and provide opportunities for hands-on participation by the public.*
The polar regions are inherently exotic to many people—the terrain, the plants, the animals, the weather, the remoteness—and they capture our imagination. This is key to engaging the public. There will be opportunities for formal classroom programs for people of a variety of ages, and media coverage that will provide both entertainment and enjoyable science education.

Recommendation 7: The U.S. science community and agencies should participate as leaders in International Polar Year 2007-2008.

• *Guide and contribute to IPY 2007-2008 activities and help to evolve the international framework, using the IPY as an opportunity to build long-lasting partnerships and cooperation across national borders.*
IPY 2007-2008 is an international effort, with more than 25 nations already committed to participate. Because of the strength of U.S. polar programs, our nation stands to play a leadership role in organizing and carrying out this ambitious program. Planning at the international level is under the auspices of two major organizations, the International Council for Science (ICSU) and the World Meteorological Organization (WMO), and the United States should lead the coordination with other countries through the ICSU and WMO to ensure the success of the IPY.

• *Continue to plan IPY 2007-2008 using an open, inclusive process.*
The initial impetus for organizing IPY 2007-2008 came from the science community, which has come together and worked diligently to identify activities of merit. This open process leverages the intellectual assets of the U.S. science community and should be continued.

• *Coordinate federal efforts to ensure a successful IPY effort, capitalizing on and supporting existing agency missions and creating new opportunities.*

International polar science efforts that have already been planned by the U.S. science community provide models for interagency collaboration, and additional future interagency efforts are encouraged, including coordination with the Arctic Council.

- *Continue planning for IPY 2007-2008, moving toward the creation of a more detailed science implementation plan.*

The next phase of IPY planning will need to provide concrete guidance that defines the science goals and addresses logistics and other key aspects of implementation. This phase of planning should include active participation by the U.S. science community and U.S. funding agencies and also continued efforts to coordinate with international planning activities so that resources are leveraged.

- *Provide mechanisms for individuals, early-career researchers, and small teams to contribute to the IPY.*

The overarching science goals of the IPY are broad and focused on international cooperation, but mechanisms for early-career researchers and small teams must be included in the larger IPY framework.

References

ACIA (Arctic Climate Impacts Assessment) Secretariat. 2004. Arctic Climate Impacts Assessment. Cambridge: Cambridge University Press.

Albert, M. 2004. The International Polar Year. Science 303:1437.

Alley, R. B. 2000. The Two-Mile Time Machine. Princeton, N.J.: Princeton University Press. 240 pp.

Alley, R. B. D. A. Meese, C. A. Shuman, A. J. Gow, K. C. Taylor, P. M. Grootes, J. W. C. White, M. Ram, E. D. Waddington, P. A. Mayewski, and G. A. Zielinski. 1993. Abrupt increase in Greenland snow accumulation at the end of the Younger Dryas event. Nature 362:527-529.

AMAP (Arctic Monitoring and Assessment Programme). 2002. Arctic Pollution 2002: Persistent Organic Pollutants, Heavy Metals, Radioactivity, Human Health, Changing Pathways. Oslo, Norway. AMAP. xii+112 pp.

Anandakrishnan, S., D. D. Blankenship, R. B. Alley, and P. L. Stoffa. 1998. Influence of subglacial geology on the position of a West Antarctic ice stream from seismic observations. Nature 394:62-65.

Arendt, A. A., K. A. Echelmeyer, W. D. Harrison, C. S. Lingle and V. B. Valentine. 2002. Rapid wastage of Alaska glaciers and their contribution to rising sea level. Science 297:382-386.

Baker, D. N., C. Barton, A. Rodger, B. Fraser, B. Thompson, and V. Papitashvili. 2004. Moving beyond the IGY: The electronic Geophysical Year (eGY) Concept. EOS (in press).

Barrett, P. 2003. Cooling a continent. Nature 421:221-223.

Barrie, L. A. 1986. Arctic air pollution: An overview of current knowledge. Atmospheric Environment 20:643-663.

Becker, L. B., M. L. Weisfeldt, M. H. Weil, T. Budinger, J. Carvico, K. Kern, G. Nichol, I. Shechter, R. Traystman, C. Webb, H. Wiedemann, R. Wise, G. Sopko. 2002. The PULSE initiative: Scientific priorities and strategic planning for resuscitation research and life saving therapies. Circulation 105:2562-2570.

Bell, R. E., D. D. Blankenship, C. A. Finn, D. L. Morse, T. A. Scambos, J. M. Brozena, and S. M. Hodge. 1998. Influence of sub-glacial geology on the onset of a West Antarctic ice stream from aerogeophysical observations. Nature 394:58-62.

Blankenship, D. D., R. E. Bell, S. M. Hodge, J. M. Brozena, J. C. Behrendt, and C. A. Finn. 1993. Active volcanism beneath the West Antarctic Ice Sheet and implications for ice-sheet stability. Nature 361:526-529.

Bottenheim, J. W., A. G. Gallant, and K. A. Brice. 1986. Measurements of NO_y species and O_3 at 82° N latitude. Geophysical Research Letters 13:113–116.

Boyer, B. B. and B. M. Barnes. 1999. Molecular and metabolic aspects of mammalian hibernation. Bioscience 49:713-724.

Brodeur, R. D., H. Sugisaki, and G. L. Hunt, Jr. 2002. Increases in jellyfish biomass in the Bering Sea: implications for the ecosystem. Marine Ecology Progress Series 233:89-103.

Cavalieri, D. J., P Gloerson, C.L. Parkinson, J.C. Comiso, and H. J. Zwally. 1997. Observed hemispheric asymmetry in global sea ice changes. Science 278:1104-1106.

Comiso, J. C., J. Yang, S. Honjo, and R. A. Krishfield. 2003. Detection of change in the Arctic using satellite and in situ data. Journal of Geophysical Research 108(C12):3384.

Crowley, T. J., and G. R. North. 1996. Paleoclimatology. Oxford: Oxford University Press.

Cuffey, K. M., and G. D. Clow. 1997. Temperature, accumulation, and ice sheet elevation in central Greenland through the last deglacial transition. Journal of Geophysical Research 102(26):383-396.

DeConto, R., and D. Pollard. 2003. Rapid Cenozoic glaciation of Antarctica induced by declining atmospheric CO_2. Nature 421:245-249.

DiTullio, G. R. and R. B. Dunbar. 2003. Biogeochemistry of the Ross Sea-An Introduction. In Biogeochemical Cycles in the Ross Sea, G. R. Ditullio, and R. B. Dunbar, eds. Antarctic Research Series 87:1-3.

DiTullio, G. R., J. Grebmeier, K. R. Arrigo, M. P. Lizotte, D. H. Robinson, A. Leventer, J. Barry, M. VanWoert, and R. B. Dunbar. 2000. Rapid and early export of Phaeocystis Antarctica blooms in the Ross Sea, Antarctica. Nature 404:595-598.

Dominé, F. and P. B. Shepson. 2002. Air-snow interactions and their impact on atmospheric chemistry. Science 297:1506-1510.

Doran, P. T., C. H. Fritsen, C. P. McKay, J. C. Priscu, and E. E. Adams. 2003. Formation and character of an ancient 19m ice cover and underlying trapped brine in an "ice-sealed" east Antarctic lake. Proceedings of the National Academy of Sciences 100:26-31.

Eastman, J.T., ed. 1993. Antarctic Fish Biology: Evolution in a Unique Environment. San Diego, Calif.: Academic Press.

Fetterer, F., K. Knowles, J. C. Stroeve, M. C. Serreze, J. Maslanik, C. Oelke, and T. A. Scambos. 2004. Recent Arctic Ice Extent Minima Observed with the Sea Ice Index. Submitted to EOS.

Gradinger, R. 1995. Climate change and biological oceanography of the Arctic Ocean. Philosophical Transactions of the Royal Society of London A 352:277-286.

Grebmeier, J. M., and K. H. Dunton. 2000. Benthic Processes in the Northern Bering/Chukchi Seas: Status and Global Change. In Impacts of Changes in Sea Ice and Other Environmental Parameters in the Arctic, H.P. Huntington, ed. Marine Mammal Commission Workshop, Girdwood, Alaska, 15-17 February 2000, pp. 80-93.

Gregory, J. M., P. S. Stott, D.J. Cresswell, N.A. Rayner, and C. Gordon. 2002. Recent and future changes in Arctic sea ice simulated by the HadCM3 AOGCM. Geophysical Research Letters 29(24):2175.

Grootes, P. M., and M. Stuiver. 1997. Oxygen 18/16 variability in Greenland snow and ice with 10^{-3}- to 10^5-year time resolution. Journal of Geophysical Research 102(26):455-470.

Holland, G. J., P. J. Webster, J. A. Curry, G. Tyrell, D. Gauntlett, G. Brett, J. Becker, R. Hoag, and W. Vaglienti. 2001. The Aerosonde robotic aircraft: A new paradigm for environmental observations. Bulletin of the American Meteorological Society 82:(5)889-901.

Hunt, Jr., G. L. and P. J. Stabeno. 2002. Climate change and control of energy flow in the southeastern Bering Sea. Progress in Oceanography 55:5-22.

Hunt, G. L., Jr., P. Stabeno, G. Walters, E. Sinclair, R. D. Brodeur, J. M. Napp, and N. A. Bond. 2002. Climate change and control of the southeastern Bering Sea pelagic ecosystem. Deep Sea Research Part II. 49:5821-5853.

INSROP (International Northern Sea Route Programme). 1999. Environmental Conditions Affecting Commercial Shipping on the Northern Sea Route. INSROP Working Paper #126.

IPCC (Intergovernmental Panel on Climate Change). 1998. The Regional Impacts of Climate Change: An Assessment of Vulnerability. Cambridge: Cambridge University Press.

IPCC. 2001. Climate Change 2001: The Scientific Basis. Cambridge: Cambridge University Press.

Johannessen, O. M, L. Bengtsson, M. W. Miles, S. I. Kuzima, V. A. Semenov, G. V. Alekseev, A. P. Nagurnyi, V. F. Zakharov, L. P. Bobylev, L. H. Pettersson, K. Hasselmann, and H. P. Cattle. 2004. Arctic climate change—observed and modeled temperature and sea ice variability. Tellus (A), 56A:1-18.

Johannessen, O. M., Muench, R. D., Overland, J. E., eds. 1994. The Polar Oceans and Their Role in Shaping the Global Environment: The Nansen Centennial Volume. Washington, D.C.: American Geophysical Union.

Jones, P.D. and A. Moberg. 2003. Hemispheric and large-scale surface air temperature variations: an extensive revision and an update to 2001. Journal of Climate 16:206-223.

Kaiser. J. 2003. Warmer Ocean Could Threaten Antarctic Ice Shelves. Science 302:759.

Krajick, K. 2001. Arctic Life, On Thin Ice. Science 291:424-425.

Krupnik, I., and D. Jolly, eds. 2001. The Earth is faster now: Indigenous observations of Arctic environmental change. Fairbanks, Alaska: Arctic Research Consortium of the United States.

Lu, J. Y., W. H. Schroeder, L. A. Barrie, A. Steffen, H. E. Welch, K. Martin, L. Lockhart, R. V. Hunt, G. Boila, and A. Richter, 2001. Magnification of atmospheric mercury deposition to polar regions in springtime: the link to tropospheric ozone depletion chemistry. Geophysical Research Letters 28:3219-3222.

Manabe, S. and R. J. Stouffer, 1994. Multiple-century response of a coupled ocean-atmosphere model to an increase of atmospheric carbon dioxide. Journal of Climate 7:5-23.

McConnell, J. R., G.W. Lamorey, and M.A. Hutterli. 2002. A 250-year high-resolution record of Pb flux and crustal enrichment in central Greenland, Geophysical Research Letters 29:2130-2134.

Moore, S. E., J. M. Grebmeier, and J. R. Davis. 2003. Gray whale distribution relative to forage habitat in the northern Bering Sea: current conditions and retrospective summary. Canadian Journal of Zoology 81:734-742.

Mueller D. R., W. F. Wincent and M. O. Jeffries. 2003. Break-up of the largest Arctic ice shelf and associated loss of an epishelf lake. Geophysical Research Letters 30(20):2031. doi:10.1029/2003GL017931.

Norton, D. W., ed. 2001. Fifty More Years Below Zeros: Tributes and Mediations for the Naval Arctic Research Laboratory's First Half Century at Barrow, Alaska. Fairbanks, Alaska: University of Alaska Press.

NRC (National Research Council). 2001a. Improving the Effectiveness of U.S. Climate Modeling. Washington, D.C.: National Academy Press.

NRC. 2001b. Climate Change Science: Analysis of Some Key Questions. Washington, D.C.: National Academy Press.

NRC. 2002. Abrupt Climate Change. Washington, D.C.: The National Academies Press.

NRC. 2003a. Frontiers in Polar Biology in the Genomic Era. Washington, D.C.: The National Academies Press.

NRC. 2003b. The Sun to the Earth and Beyond A Decadal Research Strategy in Solar and Space Physics. Washington, D.C.: The National Academies Press.

NRC. 2004a. Climate Data Records from Environmental Satellites Interim Report. Washington, D.C.: The National Academies Press.

NRC. 2004b. Exploration of the Seas: Voyage into the Unknown. Washington, D.C.: The National Academies Press.

NSB (National Science Board). 1987. The Role of the National Science Foundation in Polar Regions. Arlington, Va.: National Science Foundation.

NSF (National Science Foundation). 1997. The United States in Antarctica: Report of the U.S. Antarctic Program External Panel. Arlington, Va.: National Science Foundation.

Oltmans, S. J. 1981. Surface ozone measurements in clean air. Journal of Geophysical Research, 86:11,980.

ONR (Office of Naval Research), Naval Ice Center, Oceanographer of the Navy, and the Arctic Research Commission. 2001. Naval Operations in an Ice Free Arctic Symposium Final Report. Suitland, Md.: Whitney, Bradley & Brown, Inc.

Overpeck, J., K. Hughen, D. Hardy, R. Bradley, R. Case, M. Douglas, B. Finney, K. Gajewski, G. Jacoby, A. Jennings, S. Lamoureux, A. Lasca, G. MacDonald, J. Moore, M. Retelle, S. Smith, A. Wolfe, and G. Zielinski. 1997. Arctic Environmental Change of the Last Four Centuries. Science 278(5341):1251.

Price, P. B. 2000. A habitat for psychrophiles in deep Antarctic ice. Proceedings of the National Academy of Sciences, 97(3): 1247-1251.

Priscu, J. C. 2002. Commentary: Subglacial Lakes have changed our view of Antarctica. Antarctic Science. 14:291.

Priscu, J. C., E. E. Adams, H. W. Pearl, C. H. Fritsen, J. E. Dore, J. T. Lisle, C. F. Wolf, and J. A. Mikucki. 2002. Perennial Antarctic Lake Ice: A refuge for Cyanobacteria in an extreme environment. In Life in Ancient Ice, S. Rogers and J. Castello, eds. Princeton, N.J.: Princeton Press.

Putkonen, J. K. and G. H. Roe. 2003. Rain-on-snow events impact soil temperatures and affect ungulate survival. Geophysical Research Letters 30(4):1188.

Rahmstorf, S. 2003. Timing of abrupt climate change: A precise clock. Geophysical Research Letters 30:10.1029/2003GL017115.

Rapley, C. and R. Bell. 2004. Report to the ICSU Executive Board from the International Polar Year Planning Group. Available at http://www.us-ipy.org/. Accessed May 13, 2004.

Ritzwoller, M. H., N. M. Shapiro, A. L. Levshin, and G. M. Leahy. 2001. The structure of the crust and upper mantle beneath Antarctica and the surrounding oceans. Journal of Geophysical Research, 106(B12):30645-30670.

Sarmiento, J. L., T. M. C. Hughes, R. J. Stouffer, and S. Manabe. 1998. Simulated response of the ocean carbon cycle to anthropogenic climate warming. Nature 393:245-249.

Scambos, T. A., C. Hulbe, M. Fahnestock, J. Bohlander. 2000. The link between climate warming and break-up of ice shelves in the Antarctic Peninsula. Journal of Glaciology 46(154):516-30.

Schell, D. M. 2000. Declining carrying capacity in the Bering Sea: Isotopic evidence from baleen. Limnology and Oceanography 45:459-462.

SEARCH (Study of Environmental Arctic Change) Science Steering Committee. 2001. SEARCH: Study of Environmental Arctic Change, Science Plan. Seattle, Wash.: Polar Science Center, Applied Physics Laboratory, University of Washington.

Serreze, M. C. J. A. Maslanik, T. A. Scambos, F. Fetterer, J. Stroeve, K. Knowles, C. Fowler, S. Drobot, R. G. Barry, and T. M. Haran. 2003. A record minimum arctic sea ice extent and area in 2002. Geophysical Research Letters 30:1110.

Serreze, M. C., J. E. Walsh, F. S. Chapin, T. Osterkamp, M. Dyurgerov, V. Romanovsky, W. C. Oechel, J. Morison, T. Zhang, and R. G. Barry. 2000. Observational Evidence of Recent Change in the Northern High-Latitude Environment. Climatic Change 46(1-2):159-207.

Severinghaus, J. P., and E. J. Brook. 1998. Abrupt climate change at the end of the last glacial period inferred from trapped air in polar ice. Science 286:930-934.

Shapiro, N. M. and M. H. Ritzwoller. 2002. Monte-Carlo inversion for a global shear velocity model of the crust and upper mantle. Geophysical Journal International 151:88-105.

Shepherd, A., D. Wingham, T. Payne, and P. Skvarca. 2003. Larsen Ice Shelf has progressively thinned. Science 302:856-859.

Shepson, P., P. Matrai, L. Barrie, and J. Bottenheim. 2003. Ocean-Atmosphere-Sea Ice-Snowpack Interactions in the Arctic and Global Change. EOS Transactions 84:349.

Siegert, M. J., J. C. Ellis-Evans, M. Tranter, C. Mayer, J. R. Petit, A. Salamatin, and J. C. Priscu. 2001. Physical, chemical and biological processes in Lake Vostok and other Antarctic subglacial lakes. Nature 414:603-609.

Siple, P. 1931. A Boy Scout with Byrd. New York: G. P. Putnam's Sons.

Siple, P. 1959. 90 Degrees South. New York: G. P. Putnam's Sons.

Skvarca, P., W. Rack, H. Rott and T. Ibarzábal y Donángelo. 1999. Climatic trend and the retreat and disintegration of ice shelves on the Antarctic Peninsula: an overview. Polar Research 18(2):151-157.

Stone, R., and G. Vogel. 2004. A year to remember at the ends of the Earth. Science 303:1458-1461.

Thompson, D. W. J. and S. Solomon. 2002. Interpretation of recent Southern Hemisphere climate change. Science 296:895-899.

USGCRP (United States Global Change Research Program). 2001. Climate Change Impacts on the United States: The Potential Consequences of Climate Variability and Change. Cambridge: Cambridge University Press. Available at http://www.usgcrp.gov/usgcrp/nacc/default.htm. Accessed May 14, 2004.

Whitfield, J. 2003. Too hot to handle. Nature 425:338-339.

Zwiers, F. W., and H. von Storch, 2004: On the role of statistics in climate research. International Journal of Climatology, in press.

APPENDIX A

International Partners in IPY 2007-2008

From its earliest planning, the International Polar Year (IPY) 2007-2008 has evolved out of discussions among scientists from many nations. The first significant momentum took hold once the International Council for Science (ICSU) encouraged scientists to form a planning group in the summer of 2003. This group, listed below, developed the first overall guidance defining the nature and scope of IPY (e.g., Box A1), and its February 2004 report (Rapley and Bell, 2004) to the ICSU Executive Council gained the organization's official endorsement for the IPY concept. In that report, IPY 2007-2008 is envisioned to be an international program of coordinated interdisciplinary scientific research and observations in the Earth's polar regions to explore new frontiers; deepen our understanding of polar processes and their global linkages; increase our ability to detect changes; attract and develop the next generation of polar scientists, engineers, and leaders; and capture the interest of the public and decision makers.

Members of the ICSU IPY Planning Group as of spring 2004 include:

Professor Chris Rapley (Chair), United Kingdom
Dr. Robin Bell (Vice Chair), United States of America
Dr. Ian Allison, Australia
Dr. Robert Bindschadler, United States of America
Dr. Gino Casassa, Rogazinski, Chile
Professor Steve Chown, Republic of South Africa
Professor Gerard Duhaime, Canada
Professor Vladimir Kotlyakov, Russia
Professor Olav Orheim, Norway
Dr. Hanne Petersen, Denmark
Professor Dr. Zhanhai Zhang, China
Professor Michael Kuhn, Austria (IUGG liaison)
Dr. Henk Schalke, The Netherlands (IUGS liaison)

BOX A1
Suggested Core Characteristics of IPY Activities

According to the ICSU IPY Planning Group, the following are the suggested core characteristics of IPY Activities:

- High scientific quality, address important issues
- Capable of resulting in major progress
- Address one or both polar regions
- Contribute to international collaboration
- Logistically and technically feasible
- Avoid duplication or disruption of existing activities and programs
- Provide open and timely access to data
- Maximize use of logistical assets
- Address roles for young scientists
- Include outreach activities

In addition, desired but not mandatory characteristics include:

- Build on existing to add value
- Interdisciplinary
- Provide access to field sites
- Address training/capacity building
- Opportunities for regional scholarship within broader international activities
- Readily communicable to the public

Parallel with the ICSU endorsement, the World Meteorological Organization issued an endorsement of IPY, under the leadership of Russian and other scientists. With these two critical endorsements as a foundation, planning efforts have gained energy and numerous other organizations have begun developing plans for participating. Organizations and programs supporting IPY 2007-2008 as of spring 2004 include:

- Antarctic Treaty Consultative Meeting
- Arctic Climate Impact Assessment
- Arctic Council
- Arctic Ocean Science Board
- Arctic-SubArctic Ocean Flux Study
- Committee of Managers of National Antarctic Programmes
- European Polar Board
- European Space Agency
- Forum of Arctic Research Operators
- International Arctic Science Committee
- International Permafrost Association
- Intergovernmental Oceanographic Commission
- Scientific Committee for Antarctic Research
- United States Polar Research Board
- World Meteorological Organization

In response to a call from the ICSU IPY Planning Group, individual nations have expressed interest in participating, with many already having formed national committees to serve as the focal point for planning and communications. Some of the nations that have expressed intent to participate include:

- Argentina
- Australia
- Austria
- Belgium
- Canada
- Chile
- China
- Denmark
- Finland
- France
- Germany
- India
- Italy
- Ireland
- Japan
- Russia
- South Africa
- Sweden
- Switzerland
- Netherlands
- New Zealand
- Norway
- United Kingdom
- United States of America

Appendix B

Biographical Sketches of Committee Members

Mary Albert (Chair) is a senior research scientist at the U.S. Army Engineer Research and Development Center Cold Regions Research and Engineering Laboratory. She is also adjunct professor at the Thayer School of Engineering and the Environmental Sciences Department at Dartmouth College. Her research interests include flow and transport in porous media, surface-air physical and chemical exchange processes, snow physics, numerical modeling, effects of postdepositional processes in snow and firn on ice core interpretation, and impacts of snow photochemistry on atmospheric composition. She has spent many field seasons conducting research in the deep field in Greenland and Antarctica and is a member of the National Research Council's Polar Research Board. Dr. Albert earned her Ph.D. in applied mechanics and engineering sciences in 1991 from the University of California, San Diego.

Robert Bindschadler is a senior fellow at the National Aeronautics and Space Administration's Goddard Space Flight Center, a fellow of the American Geophysical Union, and a past president of the International Glaciological Society. He maintains an active interest in the dynamics of glaciers and ice sheets, primarily on Earth, investigating how remote sensing can be used to improve our understanding of the role of ice, in the Earth's climate; dynamics of glaciers and ice sheets; understanding the role of ice in the Earth's climate; glaciological research, including measuring ice velocity and elevation, monitoring ice sheet melt, and detecting changes in ice sheet volume. Dr. Bindschadler earned his Ph.D. in geophysics from the University of Washington in 1978.

Cecilia Bitz is a physicist at the Polar Science Center, University of Washington. Her research interests include climate dynamics, high-latitude climate variability, Arctic/North Atlantic interaction, polar amplification, global coupled climate modeling, paleoclimate, climate change, and sea ice model development. Dr. Bitz earned her Ph.D. in atmospheric sciences from the University of Washington.

Jerry Bowen is a senior national correspondent for CBS News based in Los Angeles. During his 28-year network career, he has covered a broad range of environmental and science stories—from debt for nature swaps to preserve rain forests in Costa Rica to the effects of methane gas drilling in the American West to investigations of climate change in the polar regions to the impact of the Exxon Valdez oil spill. He is a veteran of Antarctic and Arctic assignments, having reported at length from Antarctica in 1991 and 1999 and from the Arctic Ocean in 1998 on Ice Station SHEBA, The Surface Heat Budget of the Arctic Ocean, and more recently in 2002 aboard the U.S. Coast Guard icebreaker *Healy* with scientists collecting data for the Western Arctic Shelf-Basin Interactions project. Mr. Bowen is a 1973 graduate of the University of Minnesota with a B.A. in journalism.

David Bromwich is a senior research scientist and director of the Polar Meteorology Group at the Byrd Polar Research Center of Ohio State University. He is also a professor with the Atmospheric Sciences Program of the Department of Geography. Dr. Bromwich's research interests include: the climatic impacts of the Greenland and Antarctic ice sheets; coupled mesoscale-global circulation model simulations; the atmospheric moisture budget of high southern latitudes, Greenland, and the Arctic basin using numerical analyses; and the influence of tropical ocean-atmosphere variability on the polar regions. Dr. Bromwich has served on the National Research Council's Committee on Geophysical and Environmental Data and was previously a U.S. Representative of the Scientific Committee on Antarctic Research. He is a member of the American Meteorological Society, the American Geophysical Union, the Royal Meteorological Society, and the American Association of Geographers. Dr. Bromwich earned his Ph.D. in meteorology from the University of Wisconsin, Madison, in 1979.

Richard Glenn is the vice president of lands for the Arctic Slope Regional Corporation. His professional experience includes petroleum geological studies, field geological mapping, structural geological and seismic interpretation, permafrost, methane hydrate, and borehole temperature profile research. Other specialties include year-round studies of the physical properties of sea ice near Barrow, Alaska; and temperature, salinity and crystallographic profiles of first- and multiyear sea ice and documentation of freeze-up, ice movement events, and spring thaw. He has served as director of the Department of Energy Management, North Slope Borough; general manager of Barrow Technical Services, a technical firm that provided project management consulting and geological and scientific research support services; and a geologist for the Arctic Slope Consulting Group. Mr. Glenn is a member of the Ilisagvik College Board of Trustees, board president of the Barrow Arctic Science Consortium, and former member of the U.S. Arctic Research Commission. Mr. Glenn is an Alaska native and earned his master's in geology from the University of Fairbanks.

Jacqueline Grebmeier is a research professor at the University of Tennessee. Her research interests include pelagic-benthic coupling, benthic carbon cycling, and benthic faunal population structure in the marine environment; understanding how water column processes influence biological productivity in Arctic waters and sediments and how materials are exchanged between the sea bed and overlying waters; and documenting longer-term trends in the ecosystem health of Arctic continental shelves. Some of her research includes analyses of the importance of benthic organisms to higher levels of the Arctic food web, including walruses, gray whale, and diving sea ducks, and studies of radionuclide distributions of sediments within the water column in the Arctic as a whole. Dr. Grebmeier earned her Ph.D. in biological oceanography in 1987 from the University of Alaska, Fairbanks.

John Kelley is a professor of marine science at the University of Alaska, Fairbanks. His research interests include geophysics and geochemistry, with emphasis on micrometeorology and trace gas processes; applied oceanography, marine and riverine acoustics, and environmental radioactivity and contaminants; and collaborative research on ice engineering and ice core drilling. Dr. Kelley earned his Ph.D. in chemical oceanography from Nagoya University, Japan, in 1974.

Igor Krupnik is an ethnologist/research anthropologist with the Arctic Studies Center of the Smithsonian Institution. Dr. Krupnik was born in Russia, where he trained as a geographer and cultural anthropologist. He has degrees in Geography (from Moscow State University), ethnography/cultural anthropology (Ph.D., 1977, Institute of Ethnology, Russian Academy of Sciences), and in ecology/subsistence management (full doctorate, 1991, Institute of Ecology, Russian Academy of Sciences). His primary research fields are modern cultures, ecology and subsistence economies of the people of the Arctic. Dr. Krupnik has done extensive field studies in the western Arctic, primarily in Alaska (St. Lawrence Island) and the Bering Strait area (since 1971) and also along the Russian Arctic coast.

Louis Lanzerotti is a distinguished member of the technical staff (emeritus) of Bell Laboratories, Lucent Technologies, and distinguished research professor at the New Jersey Institute of Technology. His principal research interests have included space plasmas, geophysics, and engineering problems related to the impact of space processes on space and terrestrial technologies. He has been a coinvestigator and principal investigator on several National Aeronautics and Space Administration (NASA) missions, including Galileo and Ulysses, and has conducted extensive ground-based and laboratory research on space and geophysics topics. He was chair (1984-1988) of NASA's Space and Earth Science Advisory Committee and a member of the 1990 Advisory Committee on the Future of the U.S. Space Program. He has also served as chair (1988-1994) of the Space Studies Board of the National Research Council (NRC) and as a member (1991-1993) of the Vice President's Space Policy Advisory Board. He served on the NRC's Polar Research Board (1982-1990) and was chair of the NRC Committee on Antarctic Science and Policy (1992-1993). Dr. Lanzerotti is a member of the International Academy of Astronautics and the National Academy of Engineering and is a fellow of the Institute of Electrical and Electronics Engineers, the American Geophysical Union, the American Institute of Aeronautics and Astronautics, the American Physical Society, and the American Association for the Advancement of Science.

Peter Schlosser is the Vinton Professor of Earth and Environmental Engineering and professor of Earth and Environmental Sciences at Columbia University and senior research scientist at the Lamont-Doherty Earth Observatory. He also is the associate director of the Earth Institute at Columbia University. He received his Ph.D. in Physics at the University of Heidelberg, Germany, in 1985. Dr. Schlosser's research interests include studies of water movement and its variability in natural systems (oceans, lakes, rivers, groundwater) using natural and anthropogenic trace substances and isotopes as "dyes" or as "radioactive clocks"; ocean/atmosphere gas exchange; reconstruction of continental paleotemperature records using groundwater as an archive; and anthropogenic impacts on natural systems. He participated in seven major ocean expeditions, five to the polar regions. He was or presently is a member or chair of national and international science steering committees, including the World Ocean Circulation Experiment, the Climate Variability and Predictability Experiment, the World Climate Research Program, the Surface Ocean Lower Atmosphere Study, and the Study of Environmental Arctic Change.

Philip M. Smith is a partner in the consulting firm of McGeary & Smith. As an organization executive, chair or member of advisory committees, and a science and technology policy consultant, he is a leader in developing effective national and international science and technology policies and an expert in theory and practice of providing scientific advice to governments and international organizations. Mr. Smith was executive officer of the National Research Council for 13 years. He previously held senior positions in the White House Office of Science and Technology Policy, the Office of Management and Budget, and the National Science Foundation. He participated in the International Geophysical Year (IGY) and was involved in the organization and management of the U.S. Antarctic Program that followed the IGY. He served on several recent NRC committees, which reviewed the science, technology, and health aspects of the foreign policy agenda of the United States, the science advisory mechanisms of the United Nations system, and the role of science and technology in countering terrorism. Dr. Smith led a review of the mission, organization, and operating practices of the Scientific Committee on Antarctic Research and, with Michael McGeary, evaluated the organization and function of seven U.S. national committees for the international unions in the mathematical and physical sciences of the International Council for Science. He was awarded a D.Sc. (honoris causa) by North Carolina State University in recognition of his public service in science and technology policy.

George Somero is the David and Lucile Packard Professor of Marine Science and director of the Hopkins Marine Station of Stanford University. His research centers on the physiological, biochemical, and molecular mechanisms used by organisms to adapt to environmental variation, notably in temperature and ambient salinity. Current studies focus on amino acid substitutions that are important in the adaptation of proteins to temperature, physiological determinants of biogeographic patterning, the physiology of invasive species, and the effects of environmental variation on gene expression. He previously served as a member on the National Research Council's Committee on Frontiers in Polar Biology. Dr. Somero is a member of the National Academy of Sciences. He earned his Ph.D. in biological sciences in 1967 from Stanford University, conducting research on Antarctic fishes.

Cristina Takacs-Vesbach is an assistant professor at the University of New Mexico. Dr. Takacs-Vesbach's research interests include discovering the diversity of microorganisms and what determines their patterns of distribution and productivity in the natural environment, bacterioplankton dynamics in Antarctica; and responses of bacterial growth to inorganic and organic nutrients. She earned her Ph.D. in microbial ecology from Montana State University.

Gunter Weller is professor of geophysics emeritus and director of the Center for Global Change Research and the National Oceanic and Atmospheric Administration (NOAA)-University of Alaska Cooperative Institute for Arctic Research in Fairbanks, Alaska, as well as executive director of the international Arctic Climate Impact Assessment. He arrived in Alaska in 1968 from Australia, where he earned his Ph.D. in Antarctic meteorology and glaciology. Both his previous research in Antarctica and his studies in the Arctic were on microclimates and on climate change and its impacts in the polar regions. He was involved in Arctic science planning and management as program manager of the National Science Foundation's polar programs in meteorology (1972-1974), project manager of the NOAA-Bureau of Land Management Outer Continental Shelf Environmental Assessment Program in the Arctic (1975-1982), project director of the National Aeronautics and Space Administration-University of Alaska Synthetic Aperture Radar Facility (1986-1993), and deputy director of the Geophysical Institute (1984-1986) and (1990-1997). Among many scientific committee assignments he was the president of the International Council for Science's International Commission on Polar Meteorology (1980-1983) and chaired the National Research Council's Polar Research Board (1985-1990).

Douglas Wiens is a professor of earth and planetary sciences at Washington University in St. Louis. His research interests include the structure of island arcs and oceanic spreading centers, anisotropy and flow patterns in the mantle, and the crustal and upper-mantle structure of Antarctica. He has directed field instrumentation programs in the Antarctic Peninsula and Trans-Antarctic Mountains. Dr. Wiens has served on the executive committee of the Incorporated Research Institutions in Seismology, the RIDGE and MARGINS steering committees, the Ocean Bottom Seismograph Instrumentation Pool oversight committee (as chair), and the Ocean Drilling Program Science Committee. He earned his Ph.D. in geological sciences from Northwestern University in 1985.

Ex officio members[1]

Mahlon C. Kennicutt II is director of the Geochemical and Environmental Research Group, professor of oceanography, member of the toxicology faculty, and team leader of the Sustainable Coastal Margins Program at Texas A&M University. His research interests include environmental monitoring, the fate and effects of contaminants, environmental impacts of offshore energy exploration and exploitation, coordination of the social and physical sciences to address environmental issues, and all aspects of the sustainable development of coastal margins. He is currently an ex officio member of the National Research Council's Polar Research Board and U.S. delegate to the Scientific Committee on Antarctic Research. Previously he served on the NRC's Committee to Review the Oil Spill Recovery Institute and the Committee on Cumulative Environmental Effects of Oil and Gas Activities on Alaska's North Slope. Dr. Kennicutt is a

---

[1]These members of the Polar Research Board served in an ex officio capacity to provide liaisons to the International Council for Science International Polar Year Planning Group, the Scientific Committee on Antarctic Research, and the International Arctic Science Committee.

member of various professional organizations, including the American Geophysical Union, the American Association for the Advancement of Science, and the American Society of Limnology and Oceanography. Dr. Kennicutt earned his Ph.D. in oceanography in 1980 from Texas A&M University.

Robin Bell is a Doherty senior research scientist at the Lamont-Doherty Earth Observatory of Columbia University where she directs research programs on the Hudson River and in Antarctica. She is a geophysicist who earned her Ph.D. in 1989 from Columbia University. Her research interests are in linking the Earth's physical processes with the impacts on biota. These interests range from linking glacial and tectonic processes to subglacial ecosystems, to understanding the ecosystem services provided to humans by rivers, estuaries, and coastal environments. She is currently the U.S. representative to the Working Group on Geophysics of the Scientific Committee on Antarctic Research and vice-chair of the International Council for Science Planning Group for the International Polar Year.

Patrick J. Webber is a professor of plant biology and director of the Arctic Ecology Laboratory at Michigan State University. His research interests span many aspects of global change. In particular, he studies the response of plants and vegetation to climate, land use, land cover, and social change. He is active in promoting the development of a pan-arctic network of environmental observatories. He is an ex officio member of the National Research Council's Polar Research Board and the U.S. delegate to the International Arctic Science Committee (IASC). Dr. Webber has been a fellow of the Arctic Institute of North America since 1978 and president of International Arctic Science Committee since 2002.

Terry Wilson is an associate professor of geological sciences at Ohio State University. Her research interests include understanding intraplate neotectonic processes in Antarctica and mid-continent North America; structural kinematic analysis of faulting in rifts and thrust belts; crustal stress determinations from boreholes and volcanic alignments; Global Positioning System measurement of crustal motions; and structural interpretation of satellite imagery, seismic profiles, and airborne magnetic data. Dr. Wilson is the U.S. alternate delegate to The Scientific Committee on Antarctic Research, chair of SCAR's Group of Specialists on Antarctic Neotectonics, and has extensive experience working to create and sustain international programs and collaboration. Dr. Wilson earned her Ph.D. in geology from Columbia University in 1983.

NRC STAFF

Sheldon Drobot has been a program officer at the Polar Research Board and the Board on Atmospheric Sciences and Climate since December 2002. He received his Ph.D. in geosciences (climatology specialty) from the University of Nebraska, Lincoln. Dr. Drobot has directed National Research Council studies that produced the reports *Elements of a Science Plan for the North Pacific Research Board* (2004) and *Climate Data Records from Environmental Satellites* (2004). His research interests include sea ice-atmosphere interactions, microwave remote sensing, statistics, and long-range climate outlooks. Dr. Drobot currently is researching interannual variability and trends in Arctic sea ice conditions and how low-frequency atmospheric circulation affects sea ice distribution, short-range forecasting of Great Lakes ice conditions, and biological implications of sea ice variability.

Chris Elfring is director of the Polar Research Board (PRB) and Board on Atmospheric Sciences and Climate (BASC). She is responsible for all aspects of strategic planning, project development and oversight, financial management, and personnel for both units. Since joining the PRB in 1996, Ms. Elfring has overseen or directed studies that produced the following reports: *Frontiers in Polar Biology in the Genomics Era* (2003), *Cumulative Environmental Impacts of Oil and Gas Activities on Alaska's North Slope* (2003), *A Century of Ecosystem Science: Planning Long-term Research in the Gulf of Alaska* (2002), and *Enhancing NASA's Contributions to Polar Science* (2001). In addition, she is responsible for the Board's activities as the U.S. National Committee to the Scientific Committee on Antarctic Research.

Kristen Averyt was a Christine Mirzayan Intern at the Polar Research Board during the summer of 2003. She received her M.Sc. in chemistry from the University of Otago (New Zealand) in 1999 and will complete her Ph.D. in geological and environmental science at Stanford University in December 2004. Her research interests include paleoceanography and paleoclimatology, trace metal cycling and speciation in seawater, and sedimentary geochemistry.

Sarah Capote is a project assistant with the Ocean Studies Board. She earned a B.A. in history from the University of Wisconsin, Madison, in 2001. Ms. Capote has worked on the following reports: *Exploration of the Seas: Voyage into the Unknown* (2003), *Nonnative Oysters in the Chesapeake Bay* (2004), *Elements of a Science Plan for the North Pacific Research Board* (2004), and *Future Needs in Deep Submergence Science: Occupied and Unoccupied Vehicles in Basic Ocean Research* (2004).

Rachael Shiflett is a senior project assistant with the Polar Research Board. She received her M.Sc. in environmental law from Vermont Law School in 2001 and will complete her J.D. at Catholic University in May 2007. Her research interests include the Endangered Species Act, the Marine Mammal Protection Act, and the National Environmental Policy Act.

Appendix C

Acronyms and Abbreviations

AIM	Aeronomy of Ice in the Mesosphere
AOSB	Arctic Ocean Studies Board
AUV	autonomous underwater vehicle
CALIPSO	Cloud-Aerosol Lidar and Infrared Pathfinder Satellite Observations
EOS	Earth Observing System
IASC	International Arctic Science Committee
ICSU	International Council for Science
IGY	International Geophysical Year
IPCC	International Panel on Climate Change
IPY	International Polar Year
LDCM	Landsat Data Continuity Mission
NASA	National Aeronautics and Space Administration
NOAA	National Oceanic and Atmospheric Administration
NPOESS	National Polar-orbiting Operational Environmental Satellite System
NPP	NPOESS Preparatory Project
NPR	National Public Radio
NSF	National Science Foundation
OCO	Orbiting Carbon Observatory
PRB	Polar Research Board
ROV	remotely operated vehicle

SCAR   Scientific Committee on Antarctic Research
SCOR   Scientific Committee for Oceanic Research
SDO   Solar Dynamics Observatory
ST5   Space Technology 5
STEREO   Solar-Terrestrial Relations Observatory

THEMIS   Time History of Events and Macroscale Interactions during Substorms
TWINS   Two Wide-angle Imaging Neutral-atom Spectrometers

UARCTIC   University of the Arctic
UAV   unmanned aerial vehicle
UCAR   University Corporation for Atmospheric Research

WMO   World Meteorological Organization